Machine Learning in Microservices

Productionizing microservices architecture for machine learning solutions

Mohamed Abouahmed

Omar Ahmed

BIRMINGHAM—MUMBAI

Machine Learning in Microservices

Copyright © 2023 Packt Publishing

All rights reserved. No part of this book may be reproduced, stored in a retrieval system, or transmitted in any form or by any means, without the prior written permission of the publisher, except in the case of brief quotations embedded in critical articles or reviews.

Every effort has been made in the preparation of this book to ensure the accuracy of the information presented. However, the information contained in this book is sold without warranty, either express or implied. Neither the authors, nor Packt Publishing or its dealers and distributors, will be held liable for any damages caused or alleged to have been caused directly or indirectly by this book.

Packt Publishing has endeavored to provide trademark information about all of the companies and products mentioned in this book by the appropriate use of capitals. However, Packt Publishing cannot guarantee the accuracy of this information.

Publishing Product Manager: Dinesh Chaudhary
Content Development Editor: Joseph Sunil
Technical Editor: Rahul Limbachiya
Copy Editor: Safis Editing
Project Coordinator: Farheen Fathima
Proofreader: Safis Editing
Indexer: Subalakshmi Govindhan
Production Designer: Prashant Ghare
Marketing Coordinator: Shifa Ansari, Vinishka Kalra

First published: February 2023

Production reference: 1170223

Published by Packt Publishing Ltd.
Livery Place
35 Livery Street
Birmingham
B3 2PB, UK.

ISBN 978-1-80461-774-8

www.packtpub.com

To my mother, who endured a lot to shape me into who I am today. To my wife, whose unwavering support has been crucial in my journey. And to my late father, who I deeply miss and who continues to inspire me every step of the way. I love you all, thank you for everything you have done and continue to do for me. This book is a testament to your support and a small token of my appreciation to you all.

– Mohamed Abouahmed

To my mother and father, who have been an inspiration, for pushing me to become the best version of myself. And to my brother, who has been like a great friend with his continual support and motivation. I wouldn't be the person I am today without you all in my life. Thank you!

– Omar Ahmed

Contributors

About the authors

Mohamed Abouahmed is a principal architect and consultant providing a unique combination of technical and commercial expertise to resolve complex and business-critical issues through the design and delivery of innovative, technology-driven systems and solutions. He specializes in network automation solutions and smart system development and deployment.

Mohamed's hands-on experience, underpinned by his strong project management and academic background, which includes a PMP certification, master's of global management, master of business administration (international business), master of science in computer networking, and BSc in Electronics Engineering, has helped him develop and deliver robust solutions for multiple carriers, service providers, enterprises, and Fortune 200 clients.

Omar Ahmed is a skilled computer engineer with experience at various start-ups and corporations working on a variety of projects, from building scalable enterprise systems to deploying powerful machine learning models.

He has a bachelor's degree in computer engineering from the Georgia Institute of Technology and is currently finishing up his master's in computer science with a specialization in machine learning at the Georgia Institute of Technology.

About the reviewer

Sumedh Datar is a senior machine learning engineer with more than 6 years of work experience in the fields of deep learning, machine learning, and software engineering. He has a proven track record of single-handedly delivering end-to-end engineering solutions to real-world problems. He works at the intersection of engineering, research, and product and has developed deep learning-based products from scratch that have been used by a lot of end customers. Currently, Sumedh works in R&D, where he works on applied deep learning with less data, and has been granted several patents and applied for several more. Sumedh studied biomedical engineering focused on computer vision and then went on to pursue a master's in computer science focused on AI.

Table of Contents

Part 2: Overview of Machine Learning Algorithms and Applications

6

Stabilizing the Machine Learning System   113

7

How Machine Learning and Deep Learning Help in MSA Enterprise Systems   125

Part 3: Practical Guide to Deploying Machine Learning in MSA Systems

8

The Role of DevOps in Building Intelligent MSA Enterprise Systems 137

12

Deploying, Testing, and Operating an Intelligent MSA Enterprise System

Preface

Machine Learning (ML) has revolutionized the technology industry and our daily lives in ways that were previously thought impossible. By combining ML algorithms with **Microservices Architecture** (**MSA**), organizations are able to create intelligent, robust, flexible, and scalable enterprise systems that can adapt to changing business requirements and improve overall system performance.

This book is a comprehensive guide that covers different approaches to building intelligent MSA systems and solving common practical challenges faced in system design and operations.

The first part of the book provides a comprehensive introduction to MSA and its applications. You will learn about common enterprise system architectures, the concepts and value of MSA, and how it differs from traditional enterprise systems. In this part, you will gain an understanding of the design, deployment, and operation of MSA, including the basics of DevOps processes.

The second part of the book dives into ML and its applications in MSA systems. You will learn about the key ML algorithms and their applications in MSA, including regression models, multiclass classification, text analysis, and **Deep Learning** (**DL**). This part provides a comprehensive guide on how to develop the ML model, components, and sub-components, and how to apply them in an MSA system.

The final part of the book brings together everything covered in the previous parts. It provides a step-by-step guide to designing and developing an intelligent system, with hands-on examples and actual code that can be imported for real-life use cases. You will also learn about the application of DevOps in enterprise MSA systems, including organizational structure alignment, quality assurance testing, and change management.

By the end of this book, you will have a solid understanding of MSA and its benefits and will be equipped with the skills and knowledge necessary to build your own intelligent MSA system and take the first step toward achieving better business results, operational performance, and business continuity. Whether you are a beginner or an experienced developer, this book is the perfect guide to help you understand and apply MSA and ML in your enterprise systems.

Who this book is for

This book is ideal for ML solution architects, system and ML developers, and system and solution integrators. These individuals will gain the most from this book as it covers the critical concepts and best practices in ML. The book is written to provide these professionals with the knowledge and skills necessary to implement intelligent MSA solutions.

To fully benefit from this book, you should have a basic understanding of system architecture and operations. Additionally, a working knowledge of the Python programming language is highly desired. This is because the examples and case studies included in the book are primarily implemented in Python. However, the concepts and best practices covered in this book can be applied to other programming languages and technologies as well. The book is designed to provide a solid foundation in ML, while also helping you to deepen your existing knowledge and skills.

What this book covers

Chapter 1, Importance of MSA and Machine Learning in Enterprise Systems, provides an introduction to MSA and its role in delivering competitive and reliable enterprise systems. The chapter will compare MSA with traditional monolithic enterprise systems and discuss the benefits and challenges of deploying and operating MSA systems. It will also cover the key concepts of MSA, including service-driven and event-driven architecture and the importance of embracing DevOps in building MSA systems.

Chapter 2, Refactoring Your Monolith, focuses on the transition process from a monolithic architecture to an MSA. It emphasizes how to refactor the monolithic system to build a flexible and reliable MSA system. This chapter will explore the steps necessary to transition to MSA, including identifying microservices, breaking down business requirements, and decomposing functions and data. The chapter will provide insights into how to modernize an organization's enterprise systems through MSA adoption.

Chapter 3, Solving Common MSA Enterprise System Challenges, discusses the methodologies of addressing the challenges of maintaining a reliable, durable, and smoothly operating MSA system. The chapter covers topics such as using an **Anti-Corruption Layer** (ACL) for MSA system isolation, API gateways, service catalogs and orchestrators, a microservices aggregator, and a microservices circuit breaker; the differences between gateways, orchestrators, and aggregators; and other MSA system enhancements.

Chapter 4, Key Machine Learning Algorithms and Concepts, provides a comprehensive understanding of the fundamental AI, ML, and DL concepts, to equip you with the necessary knowledge to build and deploy AI models in MSA systems. It covers the differences between these areas and provides an overview of common ML packages and libraries used in Python. The chapter then dives into various applications of ML, including building regression models, multiclass classification, text sentiment analysis and topic modeling, pattern analysis and forecasting, and building enhanced models using DL.

Chapter 5, Machine Learning System Design, provides a comprehensive understanding of the design considerations and components involved in building an ML pipeline and equips you with the knowledge necessary to build and deploy a robust and efficient ML system. The chapter covers the main concepts of fit and transform interfaces, train and serve interfaces, and orchestration.

Chapter 6, Stabilizing the Machine Learning System, arms you with a comprehensive understanding of the phenomenon of dataset shifts and how to address them in your ML systems to ensure stable and accurate results. The chapter discusses optimization methods that can be applied to address dataset shifts while maintaining their functional goals. The chapter covers details on the concepts of ML parameterization, the causes of dataset shifts, the methods for identifying dataset shifts, and the techniques for handling and stabilizing dataset shifts.

Chapter 7, How Machine Learning and Deep Learning Help in MSA Enterprise Systems, wraps up all the previous chapters by discussing the different use cases where you can apply ML and DL to your intelligent enterprise MSA system. You will learn some possible use cases, such as pattern analysis using a supervised linear regression model and self-healing using DL.

Chapter 8, The Role of DevOps in Building Intelligent MSA Systems, teaches you how to apply the concepts of DevOps in building and running an MSA system. The chapter covers the alignment of DevOps with the organizational structure, the DevOps process in enterprise MSA system operations, and the application of DevOps from the start to operations and maintenance.

Chapter 9, Building an MSA with Docker Containers, provides an introduction to containers and their use in building a simple project using Docker, a widely used platform in the field. The chapter covers an overview of containers and their purpose, the installation of Docker, the creation of our sample project's containers, and inter-communication between microservices in the MSA project. The objective is to provide you with a comprehensive understanding of containers and how they can be utilized in the MSA.

Chapter 10, Building an Intelligent MSA System, combines the concepts of MSA and **Artificial Intelligence (AI)** to build a demo Intelligent-MSA system. The system will use various AI algorithms to enhance the performance and operations of the original MSA demo system created earlier in the book. The Intelligent-MSA will be able to detect potential problems in traffic patterns and self-rectify or self-adjust to prevent the problem from occurring. The chapter covers the advantages of using ML, building the first AI microservice, a demonstration of the Intelligent-MSA system in action, and the analysis of AI services' operations. The goal is to provide you with a comprehensive understanding of how AI can be integrated into an MSA system to enhance its performance and operations.

Chapter 11, Managing the New System's Deployment – Greenfield Versus Brownfield, introduces you to the deployment of Intelligent-MSA systems in greenfield and brownfield deployments. It provides ways to smoothly deploy the new system while maintaining overall system stability and business continuity. The chapter covers deployment strategies, the differences between greenfield and brownfield deployments, and ways to overcome deployment challenges, particularly in brownfield deployments where existing systems are already in production.

Chapter 12, Deploying, Testing, and Operating an Intelligent MSA System, is the final chapter, and it integrates all the concepts covered in the book to provide hands-on and practical examples of deploying an intelligent MSA system. It teaches you how to apply the concepts learned throughout the book to your own deployment needs and criteria. The chapter assumes a brownfield environment with an existing monolithic architecture system and covers overcoming deployment dependencies, deploying the MSA system, testing and tuning the system, and conducting a post-deployment review.

To get the most out of this book

To maximize your learning experience, you should have a basic understanding of system architecture, software development concepts, DevOps, and database systems. Previous experience with MySQL and Python is not necessary, but it will help in understanding the concepts in the code examples more efficiently.

Software/hardware covered in the book	Operating system requirements
Docker	Ubuntu Linux or macOS
Python	Linux, Windows, or macOS
MySQL	Ubuntu Linux or macOS
VirtualBox	Windows or macOS

We recommend you install a Python IDE such as PyCharm to be able to follow the Python examples. PyCharm can be downloaded from `https://www.jetbrains.com/lp/pycharm-anaconda/`.

VirtualBox is used to build the demo environment and create test virtual machines. VirtualBox can be downloaded from `https://www.virtualbox.org/wiki/Downloads`.

Ubuntu Linux is what we used in the book to install Docker and other utilities. To download the latest Ubuntu version, use the following link: `https://ubuntu.com/desktop`.

If you are using the digital version of this book, we advise you to type the code yourself or access the code from the book's GitHub repository (a link is available in the next section). Doing so will help you avoid any potential errors related to the copying and pasting of code.

Download the example code files

You can download the example code files for this book from GitHub at `https://github.com/PacktPublishing/Machine-Learning-in-Microservices`. If there's an update to the code, it will be updated in the GitHub repository.

We also have other code bundles from our rich catalog of books and videos available at `https://github.com/PacktPublishing/`. Check them out!

Conventions used

There are a number of text conventions used throughout this book.

`Code in text`: Indicates code words in text, database table names, folder names, filenames, file extensions, pathnames, dummy URLs, user input, and Twitter handles. Here is an example: "Once you are done composing your Dockerfile, you will then need to save it as `Dockerfile` to be able to use it to create the Docker image."

A block of code is set as follows:

```
import torch
model = torch.nn.Sequential( # create a single layer Neural
Network
    torch.nn.Linear(3, 1),
    torch.nn.Flatten(0, 1)
)
loss = torch.nn.MSELoss(reduction='sum')
```

When we wish to draw your attention to a particular part of a code block, the relevant lines or items are set in bold:

```
import numpy as np
from scipy import linalg
a = np.array([[1,4,2], [3,9,7], [8,5,6]])
print(linalg.det(a)) # calculate the matrix determinate
57.0
```

Any command-line input or output is written as follows:

```
$ docker --version
Docker version 20.10.18, build b40c2f6
```

Bold: Indicates a new term, an important word, or words that you see onscreen. For instance, words in menus or dialog boxes appear in **bold**. Here is an example: "Select **System info** from the **Administration** panel."

> **Tips or important notes**
> Appear like this.

Get in touch

Feedback from our readers is always welcome.

General feedback: If you have questions about any aspect of this book, email us at customercare@ packtpub.com and mention the book title in the subject of your message.

Errata: Although we have taken every care to ensure the accuracy of our content, mistakes do happen. If you have found a mistake in this book, we would be grateful if you would report this to us. Please visit www.packtpub.com/support/errata and fill in the form.

Piracy: If you come across any illegal copies of our works in any form on the internet, we would be grateful if you would provide us with the location address or website name. Please contact us at copyright@packt.com with a link to the material.

If you are interested in becoming an author: If there is a topic that you have expertise in and you are interested in either writing or contributing to a book, please visit authors.packtpub.com.

Share Your Thoughts

Once you've read *Machine Learning in Microservices*, we'd love to hear your thoughts! Scan the QR code below to go straight to the Amazon review page for this book and share your feedback.

https://packt.link/r/1-804-61774-1

Your review is important to us and the tech community and will help us make sure we're delivering excellent quality content.

Download a free PDF copy of this book

Thanks for purchasing this book!

Do you like to read on the go but are unable to carry your print books everywhere? Is your eBook purchase not compatible with the device of your choice?

Don't worry, now with every Packt book you get a DRM-free PDF version of that book at no cost.

Read anywhere, any place, on any device. Search, copy, and paste code from your favorite technical books directly into your application.

The perks don't stop there, you can get exclusive access to discounts, newsletters, and great free content in your inbox daily

Follow these simple steps to get the benefits:

1. Scan the QR code or visit the link below

https://packt.link/free-ebook/9781804617748

2. Submit your proof of purchase

3. That's it! We'll send your free PDF and other benefits to your email directly

Part 1:
Overview of Microservices Design and Architecture

We will start *Part 1* by providing a comprehensive introduction to **Microservices Architecture (MSA)** and its application in enterprise systems. We will learn about common enterprise system architectures, the concepts and value of MSA, and how it differs from traditional enterprise systems. Throughout this part, we will gain an understanding of the design, deployment, and operations of an MSA, including the basics of DevOps processes.

We will come to understand more details on the use cases in which each enterprise architecture model is best used. *Part 1* also examines how to design a basic modular, flexible, scalable, and robust MSA, including how to translate business requirements into microservices. We will acquire in-depth information about the different methodologies used for transitioning into an MSA and the pros and cons of each approach.

This part discusses the challenges of designing a true MSA and provides the tools and techniques for tackling each of these challenges. We will learn about the enterprise system components used for optimizing system modularity, testability, deployability, and operations.

Part 1 is designed to provide readers with a solid understanding of MSA, its benefits, and how to implement MSA in their own enterprise systems.

This part comprises the following chapters:

- *Chapter 1, Importance of MSA and Machine Learning in Enterprise Systems*
- *Chapter 2, Refactoring Monolith*
- *Chapter 3, Solving Common MSA Enterprise System Challenges*

1

Importance of MSA and Machine Learning in Enterprise Systems

In today's market, the competition has never been fiercer, and user requirements for IT systems are constantly increasing. To be able to keep up with customer requirements and market demands, the need for a shorter **time-to-market** (**TTM**) for IT systems has never been more important, all of which has pushed for agile deployment and the need to streamline the development process and leverage as much code reuse as possible.

Microservices architecture (**MSA**) addresses these concerns and tries to deliver a more competitive, reliable, and rapid deployment and update delivery while maintaining an efficient, stable system operation.

In this chapter, we will learn more details about how microservices help build a modern, flexible, scalable, and resilient enterprise system. The chapter will go over key concepts in MSA and discuss the common enterprise system architectures, how each architecture is different from MSA, why they are different, and what you gain or lose when you adopt one or more architectures over the others.

We will cover the following areas as we go over the chapter:

- What MSA is and why
- MSA versus monolithic enterprise systems
- Service-driven architecture, **event-driven architecture** (**EDA**), and how to incorporate that in MSA
- Challenges of deploying and operating MSA enterprise systems
- Why it is important to embrace **DevOps** in building MSA

Why microservices? Pros and cons

Microservices is often likened to MSA. MSA refers to the way in which a complex system is built from a collection of smaller applications, where each application is designed for a specific limited-scope function. These small applications (or services, or microservices) are independently developed and can be independently deployed.

Each microservice has an API interface for communicating with other microservices in the system. The way all these individual microservices are organized together forms the larger system function.

In order to understand the value of microservices and the challenges one faces in designing an MSA, it is imperative to understand how microservices communicate and interact with each other.

Microservices can communicate together in a linear or non-linear fashion. In a linear microservices pipeline, each microservice communicates with another microservice, processing data across the system in a sequential manner. The input is always passed to the first microservice, and the output is always generated by the last microservice in the system:

Figure 1.1: Linear microservices pipeline

Practically, however, most existing systems are formed using a non-linear microservices pipeline. In a non-linear microservices pipeline, data is distributed across different functions in the system. You can pass the input to any function in the system, and the output can be generated from any function in the system. You can therefore have multiple pipelines with multiple inputs, serving multiple functions and producing multiple outputs:

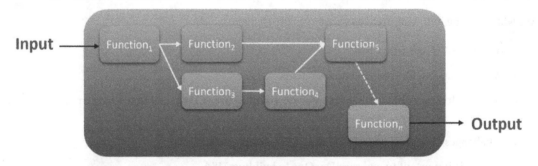

Figure 1.2: Non-linear microservices pipeline

Consider the following diagram of a simplified order fulfillment process in a typical e-commerce system. Each function within the **Placing an Order** process represents a microservice. Once an order is placed by a customer, an API call is triggered to the **Add/Update Customer Information** microservice to save that customer's information or update it if needed. This microservice sole responsibility is just that: manage customer information based on the data input it receives from the API caller.

Another API call is issued at the same time to the **Verify Payment** part of the process. The call will be directed to either the **Process PayPal Payment** or the **Process Credit Card Payment** microservice depending on the payment type of the API call. Notice here how the payment verification process is broken down into two different microservices—each is specifically designed and developed for a specific payment function. This enables the flexibility and portability of these microservices to other parts of the system or to another system if needed.

After payment is processed, API calls are triggered simultaneously to other microservices in the system to fulfill the order:

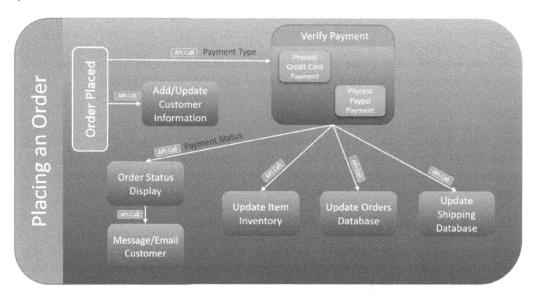

Figure 1.3: A non-linear microservices pipeline example – customer order

The order placement example shows how modular and flexible designing an MSA enterprise system can be. We will often use this example to show some of the advantages and challenges one may face when designing, deploying, and operating an MSA enterprise system.

It is essential that we go over some of the advantages and disadvantages of building enterprise systems using MSA to help decide whether MSA is a better option for your organization or not.

Note that some of the advantages listed next could also be considered disadvantages in other situations (and vice versa).

Advantages of microservices

There is some significant value to implementing MSA. The following are some of the advantages we see applicable to today's market.

Autonomy

One of the biggest advantages of microservices is their autonomy—it is the keystone for many of the other advantages of MSA. And because of their autonomy, microservices have their own technology stack, which means that each system service can be developed with completely different tools, libraries, frameworks, or programming languages than any other system service, yet they integrate with each other smoothly.

Microservices can be developed and tested independently of any other application within the system, which enables each microservice to have its own life cycle, including **quality assurance (QA)**, change management, upgrades, updates, and so on, which in return greatly minimizes application dependencies.

Portability

Microservices' autonomy enables them to be portable across platforms, operating systems, and different systems, all independent of the coding language in which these services were written.

Reuse

When reusing microservices, you don't need to reinvent the wheel. Because of their autonomy, microservices can be reused without the need to add additional coding, changes, or testing. Each service can be reused as needed, which largely increases system flexibility and scalability, significantly reduces the development time, cost, and deployment time, and reduces the system's TTM.

Loosely coupled, highly modular, flexible, and scalable

Microservices form the main building blocks of an MSA enterprise system. Each block is loosely coupled with the other blocks in the system. Just like Lego blocks, the manner in which these blocks are organized together can form a complex enterprise MSA system building a specific business solution.

The following diagram shows an example of how we can build three different systems with multiple microservices.

The diagram shows nine services, and seven out of these services are organized in such a manner to reuse and build three different systems—system **A**, system **B**, and system **C**. This shows how loose coupling enables flexibility in MSA in such a way that you can reuse each service to build a different system function.

You can build a system with minimal development added to existing microservices either acquired by a third party or previously developed in house. This largely enables rapid system development, new feature releases, very short TTM, and reliable, flexible, and much more stable hot updates and upgrades. All of this increases **business continuity** (**BC**) and makes the enterprise system much more scalable:

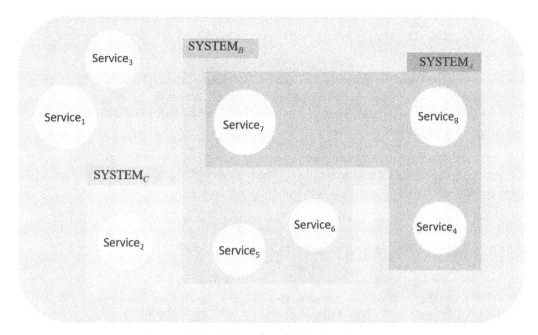

Figure 1.4: Flexibility and modularity in microservices

Shorter release cycle and TTM

Because of the individual and independent services features we previously mentioned, the deployment of microservices becomes much easier and faster to perform. Automation can play a great role in reducing time-of-service testing and deployment, as we will discuss later in this chapter.

Fault tolerance and fault isolation

Each microservice has its own separate fault domain. Failures in one microservice will be contained within that microservice, hence it is easier to troubleshoot and faster to fix and bring back the system to full operations.

Consider the order fulfillment example we mentioned earlier; the system can still be functional if the **Message/Email Customer** microservice—for example—experiences any failures. And because of the nature of the failure and the small fault domain, it will be easy to pinpoint where that failure is and how to fix it. **Mean Time to Resolution** (**MTTR**) is therefore significantly reduced, and BC is greatly enhanced.

Architects are sometimes able to build the system with high embedded tolerance to prevent these failures to begin with or have other backup microservices on standby to take over once a failure is detected in the primary microservice. One of the primary objectives of this book, as we will see later, is to be able to design a system with high enough intelligence to provide the desired high resilience.

What software architects have to bear in mind, however, is that, with too many system components in the MSA, too many things can go wrong. Architects and developers, therefore, have to have solid fallback and error handling to manage the system's resilience.

The communication between the different microservices, for example, can simply time out for whatever reason; it could be a network issue, a server issue, or too many API calls at the receiving microservices or at the event-handling mechanism developed in the system, overwhelming this system component and causing failures or delayed response.

There are many data flow streams and data processing points in the system that all need to be synchronized. A single failure, if not taken care of properly by the system, can create system-cascading failures, and accordingly could cause a failure to the entire system.

How fault tolerance is designed will be a big factor in how system performance and reliability are impacted.

Reliability and the Single Responsibility Principle (SRP)

If you come from the programming world, you are probably familiar with the SRP in **object-oriented programming** (**OOP**): *A class should have one, and only one, reason to change.* Every object, class, or function in the system should have a responsibility over only that functionality of the system, and hence that class, once developed, should only change for the reason it was originally created for. This principle is one of the main drivers of increased system reliability and BC in MSA.

At the initial phases of developing an MSA enterprise system, and during the phase of developing new microservices from scratch, the MSA enterprise system may not be fully tested or fully matured yet, and reliability may still be building up. When the system matures, changes to individual microservices are minimal—if any— and microservices' code reliability is, therefore, higher, the operation is more stable, fault domains are contained, fault tolerance is high, and the system's reliability thus becomes much higher than similar systems with a monolithic architecture. Reliability is highly contingent on how well the system is designed, developed, and deployed.

Reducing system development and operational cost

Reusing microservices largely reduces the development efforts and time needed to bring the system to life. The more microservices you can reuse, the lower the development time and cost will become.

Microservices do not have to be developed from scratch; you can purchase already developed microservices that you may need to plug into your MSA enterprise system, cutting the development time significantly.

When these microservices are stable and mature, reliability is higher, MTTR is much shorter, and hence system faults are lower and BC is higher. All these factors can play a major role in reducing the development cost, operational cost, and **total cost of ownership** (**TCO**).

Automation and operational orchestration are ideal for microservices; this enables agile development and can also decrease operational costs significantly.

Disadvantages of microservices

Microservices come with a set of challenges that need to be taken into consideration before considering an MSA in your organization. The good news is that many of these challenges—if not all—can effectively be addressed to have in the end a robust MSA enterprise system.

Mentioned here are some of the challenges of microservices, and we will later in this chapter talk about some of the methodologies that help address these challenges.

Complexity

MSA systems contain many components that must work together and communicate together to form the overall solution. The system's microservices in most cases are built with different frameworks, programming languages, and data structures.

Communication between microservices has to be in perfect synchronization for the system to properly function. Interface calls could at times overwhelm the microservice itself or the system as a whole, and therefore, system architects and developers have to continuously look for mechanisms to efficiently handle interface calls and try to eliminate dependencies as much as they can.

Designing the system to handle call loads, data flows, and data synchronization, along with the operational aspects of it, could be a very daunting process and creates layers of complexity that are hard to overlook.

Complexity is one of the main trade-off factors in implementing and running an MSA enterprise system.

Initial cost

MSA systems usually require a large number of resources to be able to handle the individual processing needs of each microservice, the high level of communication between microservices, and the different development and staging environments for developing these microservices.

If these microservices are being developed from scratch, the initial cost of building an MSA system would therefore be too high. You have to account for the cost of the many individual development environments, the many microservices to develop and test, and the different teams to do all these tasks and integrate all these components. All this adds to the cost of the initial system development.

Tight API control

Each microservice has its own API calls to be able to integrate with other microservices in the system. Any change in the API command reference set—such as updates in any API call arguments, deprecated APIs, or changes in the return values—may require a change in how other microservices handle the data flow from and to that updated microservice. This can pose a real challenge.

Developers have to either maintain backward compatibility (which can be a big constraint at times) or change the API calls' code of every other component in the system that interacts with the updated microservice.

System architects and developers have therefore to maintain very tight control over API changes in order to maintain system stability.

Data structure control and consistency

The drawback of having independent applications within the enterprise system is that each microservice will have to maintain its own data structure, which creates a challenge in maintaining data consistency across your system.

If we take the earlier example of customer order fulfillment, the **Add/Update Customer Information** microservice should have its own database totally independent from any other database in the system. Similarly, the **Update Item Inventory** microservice should be the microservice responsible for the item information database, the **Update Orders Database** microservice should have the orders database, and so on.

The challenge now is that the shipping database will need to be in sync with the customer information database, and the orders database will have to contain some of the customer information. Also, the **Message/Email Customer** microservice has to have a way to access customer information (or receive customer information through API calls), and so on. In a larger system, the process of keeping data consistent across the different microservices becomes problematic. The more microservices we have, the more complex the data synchronization becomes.

Once again, designing and developing a system with all that work in mind becomes another burden on the system architects and developers.

Performance

As we mentioned earlier, microservices have to communicate with each other to perform the entire system function. This communication, data flows, error handling, and fault-tolerance design—among many other factors—are susceptible to network latency, network congestions, network errors, application data processing time, database processing time, and data synchronization issues. All these factors greatly impact system performance.

Performance is another major trade-off factor in adopting and running an MSA enterprise system.

Security

Because of microservices' autonomy and their loose coupling, a high number of data exchanges between the different services is necessary for the MSA to function. This data flow, data storage within each microservice, data processing, the API call itself, and transaction logging all significantly increase the system attack surface and develop considerable security concerns.

Organizational culture

Each microservice in the MSA has its own development cycle and therefore has its silo of architects, developers, testers, and the entire development and release cycle teams, all to maintain the main objective of microservices: their autonomy.

MSA enterprise systems are built from a large number of microservices and mechanisms to manage the interaction between the different system components. Developers have to therefore have system operational knowledge, and the operational teams need to have development knowledge.

Testing such complex distributed environments that one will have in the MSA system becomes a very daunting process that needs a different set of expertise.

The traditional organizational structure of one big development team solely focused on development, one QA team only doing basic testing, and so on is no longer sufficient for the way MSA is structured and operated.

Agile development and DevOps methodologies are very well suited for microservices development. You need agile processes to help maintain the fast development and release cycles MSA promises to deliver. You need DevOps teams who are very familiar with the end-to-end process of designing the application itself and how it fits in the big picture, testing the application, testing how it functions within the entire system, the release cycle, and how to monitor the application post release.

All this requires a cultural shift and significant organizational transformation that can enable DevOps and agile development.

> **Important note**
>
> We rarely see a failure in MSA adoption because of technical limitations; rather, failure in adopting MSA is almost always due to a failure to shift the organization's culture toward a true DevOps and agile culture.

The benefits outweigh the detriments

The main questions you need to answer now are: Is building an MSA worth it? Can we make it happen given the current organizational culture? How long will it take the organization to transform and be ready for MSA? Do we have the luxury of waiting? Can we do both the organizational transformation and the building of the MSA enterprise system at the same time? Do we have the resources and the caliber necessary for the new organizational structure? Is cost an issue, and do I have the budget to cover that?

Well, first of all, if you are planning to build a large enterprise system, and you have the budget and necessary resources for starting this project, building the system as MSA is definitely worth it. All initial costs endured and time spent will eventually be offset by the long-term cost and time-saving benefits of having an MSA system.

Nevertheless, you are still the one to best address all these previous questions. There are overwhelming and compelling advantages to adopting MSA, but as we have seen, this is not a simple undertaking; so, whether an organization is willing to walk that path or not is something it—and only it—can answer.

Now we know what the advantages of deploying an MSA are, and the challenges that come with MSA adoption, we will now go over different enterprise architecture styles, what they are, and the differences between each other.

Loosely versus tightly coupled monolithic systems

Traditional applications back in the day were mostly built using a monolithic architecture, in which the entire application was one big code base. All system components and functions were tightly coupled together to deliver the business solution.

As shown in the following diagram, system functions are all part of the same code, tightly coupled with centralized governance. Each system function has to be developed within the same framework of the application.

In an MSA system, however, each function preserves its own anonymity—that is, loosely coupled with decentralized governance, giving each team the ability to work with its own preferred technology stack, with whichever tools, framework, and programming language it desires:

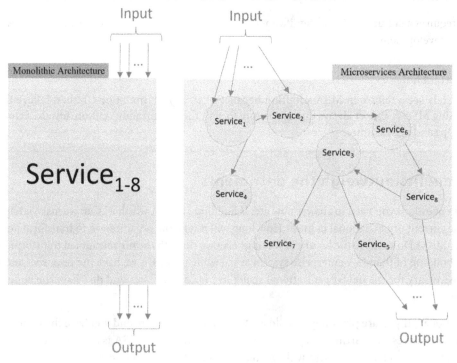

Figure 1.5: Monolithic versus microservices systems

All functions in the monolithic architecture application are wrapped into the application itself. In the MSA, these functions are developed, packaged, and deployed separately. Therefore, we can run these services in multiple locations' on-premises infrastructure, in the public cloud, or across both on-premises and the cloud in a hybrid-cloud fashion.

In monolithic systems, and because of the tight coupling, synchronizing the different system function changes is a development and operational nightmare. If one application (for whatever reason) becomes unstable, it could cause a failure to the entire system, and bringing the system back to a stable point becomes a real pain.

In the case of microservices, however, since each of these microservices is loosely coupled, changes and troubleshooting are limited to that particular microservice, as long as the microservice interface does not change.

One large piece of code, in the case of monolithic architecture, is very hard to manage and maintain. It is also hard to understand, especially in large organizations where multiple developers are working together.

In many cases such as employee turnover, for example, a developer may need to troubleshoot someone else's code, and when the application is written in a single big piece of code, things tend to be complicated, hard to trace and understand, and hard to reverse engineer and fix. Code maintenance becomes a serious problem, while in the microservices case, this humongous line of code is broken into smaller chunks of code that are easier to read, understand, troubleshoot, and fix, totally independent of the other components of the system.

When code changes are needed in monolithic architecture, a single change to part of the code may need changes to many other parts of the application, and accordingly, change updates will likely require a rewrite and a recompile of the entire application.

We can also reuse and package different applications together in a workflow to form a specific service, as shown previously in *Figure 1.4*.

It is just common sense to break down a complex application into multiple modules or microservices, each performing a specific function in the entire ecosystem for better scalability, higher portability, and more efficient development and operations.

For small, simple, and short-lived systems, monolithic applications may be a better fit for your organization, easier to design and deploy, cheaper to develop, and faster to release. As the business needs grow, MSA becomes a better long-term approach.

Since monolithic systems are tightly coupled, there is no need for API communication between the different system functions; this significantly decreases the security surface of your system, lowering system security risks and increasing the system's overall performance.

Think of the deployment difference between both monolithic and MSA as the difference between an economy car and a Boeing 787. The car is a better, cheaper, and faster tool for traveling between two cities 50 miles apart, with no need for the security checks you experience in airports before boarding your flight. As the distance increases, however, driving the car becomes more hassle. At 5,000 miles, the Boeing 787 is likely to become a better, cheaper, and faster way to get to your destination, and you will likely be willing to put up with the hassle of security checks you have to undergo to be able to board your flight.

The following is a comparison summary between both monolithic and microservices applications:

	Monolithic	MSA
Architecture	Highly autonomous. System functions are split into independent loosely coupled chunks of smaller code.	No autonomy. System functions are all tightly coupled into one big piece of code.
Portability	Highly portable	Very limited portability
Reuse	Highly reusable	Very limited ability to reuse code
Modularity and Scalability	Highly modular and scalable	Limited modularity and hard to scale
Initial TTM	Highly dependent on the readiness of individual system services. The more code reuses, the shorter the TTM is. If the system microservices are being designed and developed from scratch, TTM is usually longer for monolithic architecture.	Long TTM, especially in large systems. Shorter TTM in small and simple systems.
Release Cycle	Very short release cycle, super-fast to deploy changes and patch updates	Long and usually very time-consuming release cycles and patch updates
Initial Cost	Usually high. Depends on the system size. The initial cost is offset by operational cost savings.	Usually low. The initial size becomes higher in large enterprise systems.
Operational Cost	Low. Easier to maintain and operate.	High. Hard to maintain and operate.
Complexity	High	Low
API Control	High	Low

	Monolithic	**MSA**
Data Structure Consistency	Decentralized databases, hence data consistency is harder to maintain	A centralized database, hence easier to maintain data consistency across the system
Performance	Usually lower	Usually higher
Security	Many security concerns	Lower security concerns
Organizational Adoption	Hard to adopt depending on the organizational structure. Requires adoption of agile development and DevOps. Organizational transformation may be required and may take a long time to achieve.	Easy to adopt. Minimal organizational transformation needed—if any.
Fault Tolerance	Usually higher	Usually lower

Table 1.1: Summary of the differences between monolithic and MSA systems

We covered in this section the different aspects of a monolithic system; next, we go over service-driven architecture and EDA, and how to combine these architectural styles within MSA to address some of the MSA challenges discussed earlier.

Service-driven, EDA, and MSA hybrid model architecture

People often get mixed up between MSA and **service-driven architecture** (aka **service-oriented architecture** or **SOA**). Both types of architecture try to break down the monolithic architecture system into smaller services. However, in MSA, the system services decomposition is extremely granular, breaking down the system into very fine specialized independent services. In the SOA, the system services decomposition is instead coarse-grained to the domain level.

All domains, as shown in the following diagram, share the same centralized database and may actually share other resources in between, creating some level of coupling and system dependencies that are non-existent in MSA. Data storage is a key difference between both architectural styles:

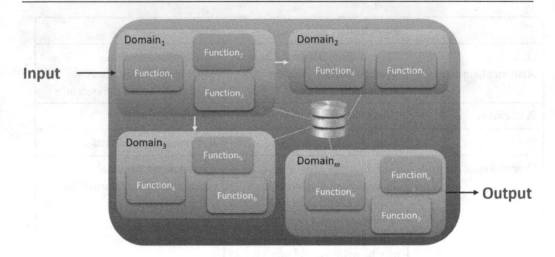

Figure 1.6: SOA architecture split into functional domains

In the case of the simplified MSA customer ordering example discussed earlier, there are eight different microservices. A similar implementation in SOA is likely to have all these microservices built together and tightly coupled in a single domain. Other domains within the system could be Cart Handling, Catalog Browsing and Suggestions, and so on.

SOA has a holistic enterprise view, while in a microservice, development looks into the function itself in total isolation of the enterprise system in which the microservice is intended to be used.

EDA is another architectural style that is largely adopted. While MSA's main focus is on function and SOA emphasizes the domain, EDA instead focuses on system events.

EDA is usually complemented by another main system architecture, such as SOA or MSA. In EDA, services are decoupled at a granularity level determined by its main architecture (MSA or SOA) and then communicate with each other through event-based transactions. In our order placement example, these events could be **Order Created**, **Order Canceled**, **Order Shipped**, and so on.

In order to maintain event synchronization and data consistency across the enterprise system, these events must be handled by a message broker. The message broker's sole responsibility is to guarantee the delivery of these events to different services across the system. Therefore, it has to be highly available, highly responsive, fault-tolerant, and scalable and must be able to function under heavy load.

When EDA is adopted within the MSA enterprise system, the message broker in that case will be handling events, API calls, and API calls' responses.

The message broker has to be able to queue messages when a specific service is down or under heavy load and deliver that message whenever that service becomes available.

ACID transactions

Any system with some form of data storage always needs to ensure the integrity, reliability, and consistency of that data. In MSA, systems store and consume data across the workflow transactions, and for individual services to ensure integrity and reliability for the MSA system as a whole, data stored within the entire system have to comply with a certain set of principles called **Atomicity, Consistency, Isolation, and Durability** (**ACID**):

- **Atomicity**: All-or-nothing transactions. Either all transactions in the workflow are successfully executed and committed or they all fail and are canceled.

- **Consistency**: Any data change in one service has to maintain its integrity across the system or be canceled.

- **Isolation**: Each data transaction has its own sovereignty and should not impact or be impacted by other transactions in the system.

- **Durability**: Committed transactions are forever permanent, even in the case of a system failure.

Saga patterns

One of the main challenges in MSA is distributed transactions, where data flow spans across multiple microservices in the system. This flow of data across the services creates a risk of violating the microservice autonomy. Data has to be managed within the microservice itself in total isolation from any other service in the system.

If you look at our order placement example again, you find that customer data (or part of it) spans across the different microservices in the example, which could create undesired dependencies in the MSA, and should be avoided at all costs.

What if, for whatever reason, the **Update Item Inventory** service fails, or it just happens that the service reports back that the item is no longer available? The system in that case will need to roll back and update all individual services' databases to ensure ACID transactions for the workflow.

The **saga pattern** manages the entire workflow of transactions. It sees all sets of transactions performed in a specific process as a workflow and ensures that all these transactions in that workflow are either successfully executed and committed or rolled back in case the workflow breaks for whatever reason, to maintain data consistency across the system.

A saga participant service would have a **local transaction** part of that workflow. A local transaction is a transaction performed within the service itself and produces an event upon execution to trigger the next local transaction in the workflow. These transactions must comply with ACID principles. If one of these local transactions fails, the saga service initiates a set of **compensating transactions** to roll back any changes caused by the already executed local transactions in the workflow.

Each local transaction should have corresponding compensating transactions to be executed to roll back actions caused by the local transaction, as shown in the following diagram:

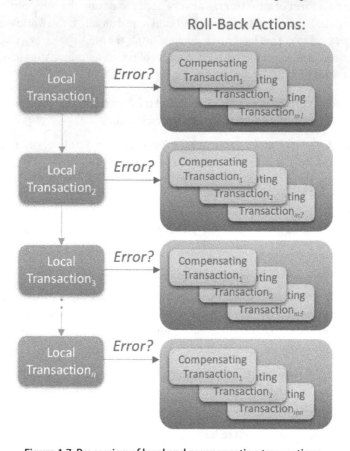

Figure 1.7: Processing of local and compensating transactions

There are two ways to coordinate transactions' workflow in a saga service: **choreography** and **orchestration**.

In choreography, saga participant services exchange events without the need for a centralized manager. As in EDA, a message broker is needed to handle event exchanges between services, as illustrated in the following diagram:

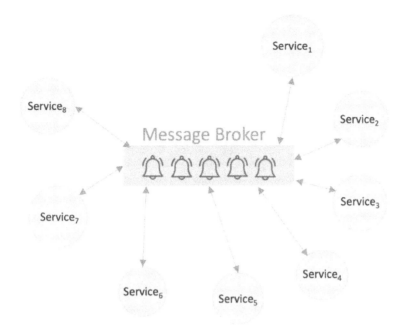

Figure 1.8: Choreography in a saga service

In orchestration, a saga pattern-centralized controller is introduced: an **orchestrator**. The workflow is configured in the orchestrator and the orchestrator sends requests to each saga participant service on which local transaction it needs to execute, receives events from saga participant services, checks the status of each request, and handles any local transaction failures by executing the necessary compensating transactions, as illustrated in the following diagram:

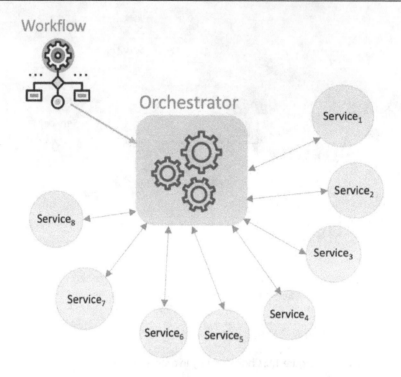

Figure 1.9: Orchestration in a saga service

Orchestrators become the brain of the enterprise system and the single source for all steps that need to be taken to execute a specific system workflow. The orchestrator, therefore, must be implemented in a way to be highly resilient and highly available.

Command Query Responsibility Segregation (CQRS)

It is very common in traditional systems, and especially in monolithic applications, to have a common relational database deployed in the backend and accessed by a frontend application. That centralized database is accessed with **Create-Read-Update-Delete (CRUD)** operations.

In modern architecture, especially as the application scales, this traditional implementation poses a problem. With multiple CRUD requests being processed on the database, table joins are created with a high likelihood of database locking happening. Table locks introduce latency and resource competition, and greatly impact overall system performance.

Complex queries have a large number of table joins and can lock the tables, preventing any write or update operations on them till the query is done and the database unlocks the tables. Database read operations are typically multiple times more than write operations, and in heavy transaction systems, the problem can multiply.

You can see a comparison of CRUD and CQRS patterns here:

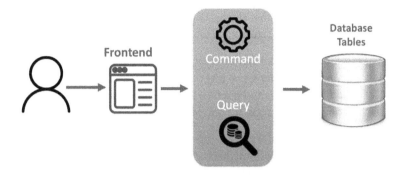

Create, Read, Update, Delete (CRUD) Model

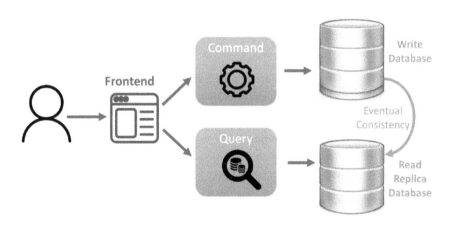

Command Query Responsibility Separation (CQRS)

Figure 1.10: CRUD versus CQRS patterns

With CQRS, you simply separate one object into two objects. So, rather than doing both commands and queries on one object, we separate that object into two objects—one for the command, and one for the query. A command is an operation that changes the state of the object, while a query does not change the state of the system but instead returns a result.

In our case, the object here is the system database, and that database separation could be either physical or logical. Although it is a best practice to have two physical databases for CQRS, you can still use the same physical database for both commands and queries. You can, for example, split the database into two logical views—one for commands and one for queries.

A replica is created from the master database when two physical databases are used in CQRS. The replica will, of course, need to be synchronized with the master for data consistency. The synchronization can be accomplished by implementing EDA where a message broker is handling all system events. The replica subscribes to the message broker, and whenever the master database publishes an event to the message broker, the replica database will synchronize that specific change.

There will be a delay between the exact time at which the master database was actually changed and when that change is reflected in the replica; the two databases are not 100% consistent during that period of time but will be eventually consistent. In CQRS, this synchronization is called eventual consistency synchronization.

When applying CQRS design in MSA, database processing latency is greatly reduced, and hence communication between individual services' performance is greatly enhanced, resulting in an overall system-enhanced performance.

The database used can be of any type, depending on the business case of that particular service in the MSA. It may very well be a **relational database (RDB)**, document database, graph database, and so on. A NoSQL database could also be an excellent choice.

We discussed previously the MSA from a design and architecture perspective. Operating the MSA system is another aspect that the entire organization must consider for a successful business delivery process. In the next section, we discuss DevOps, how it fits into the MSA life cycle, and why it is important for a successful MSA adoption and operation.

DevOps in MSA

DevOps revolves around a set of operational guidelines in the software development and release cycles. The traditional development engineer is no longer living in their confined environment where all the focus is to convert functional specifications into code; rather, they should have an end-to-end awareness of the application.

A DevOps engineer would oversee, understand, and be involved in the entire pipeline from the moment the entire application is planned out, converting business functions into code, building the application, testing it, releasing it, monitoring its operations, and coming back with the feedback necessary for enhancements and updates.

That does not necessarily mean that a DevOps engineer would be responsible for all development and operational task details. Individual responsibilities within the application team may vary in a way to guarantee a smooth **continuous integration and continuous deployment (CI/CD)** pipeline of the application:

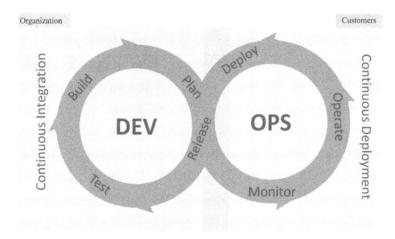

Figure 1.11: DevOps CI/CD pipeline

One of the main objectives of DevOps is to speed up the CI/CD pipeline; that's why there is a lot of emphasis on automation in DevOps. Automation is essential to efficiently perform the pipeline.

Automation can help at every step of the way. In DevOps, many test cases that are part of your QA plan are automated, which significantly speeds up the QA process. The release management and monitoring of your application are also automated to provide high visibility, continuous learning, and quick fixes whenever needed. All of this will help organizations improve productivity, predictability, and scalability.

DevOps is a holistic view of how the application is developed and managed. It is not a function for only the development team or operational team to adopt; rather, the entire organization should adopt it. It is therefore imperative for the organizational structure and the organization's vision and goal to all align with the set of procedural and functional changes necessary to shift from the traditional way of developing software.

Just to give you a gist of how traditional and DevOps models differ in terms of application development and release cycles, take a look at the following comparison table:

	Traditional	DevOps
Planning	Months Long time to plan due to the large application size and tight coupling between different application components	Days to weeks Very short planning time since the application is broken down into small individual loosely coupled services
Development	Months	Days to weeks, and even shorter in the case of patches and fixes
Testing	Weeks to months Mostly manually intensive QA use case testing, which may sometimes jeopardize the reliability of the test's outcome	Days Mostly automated QA use case execution that brings high reliability to the application
Release, Deploy	Days Usually long manual work and more susceptible to human errors	Hours Mostly automated
Operate, Monitor	Metrics reporting is mostly manually pulled and analyzed	Metrics are monitored and analyzed automatically and can even fix the problem in seconds. Moreover, **machine learning** (ML) tools can be used to enhance operations even further.

Table 1.2: Traditional operational style versus DevOps

In traditional development environments, you have a big piece of code to write, maintain, and change when needed. Because of the code size, it is only normal to have a long release cycle, and it can only be feasible to deploy patches or new releases when only major changes or high-severity fixes are needed, as illustrated in the following diagram:

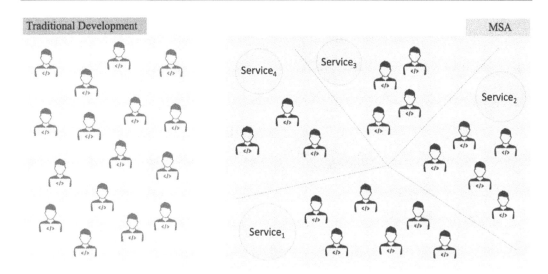

Figure 1.12: Traditional development environment versus MSA DevOps

In MSA, teams are separated based on applications that do not function. That big chunk of code is split into a collection of much smaller code (microservices), and since teams are split to work independently for each team to focus on a specialized microservice, the development and release cycles are much shorter.

Similarly, in DevOps, the application is broken down into smaller pieces to enable the CI/CD pipeline, which makes DevOps the perfect model that fits MSA.

Why ML?

Using ML tools and algorithms in your MSA enterprise system can further enhance and accelerate your DevOps CI/CD pipeline. With ML, you can find patterns in your tests, monitor phases of your pipeline, automatically analyze where the faults may be, and suggest a resolution or automatically fix operational issues whenever possible.

ML can greatly shorten your MSA enterprise system's TTM and make it more intelligent, self-healing, resilient, and supportable.

We will in this book discuss two aspects of ML: first, we'll explain in detail how to add CI/CD pipeline intelligence to your MSA enterprise system, and second, we'll look at how to build an ML enterprise system with MSA in mind:

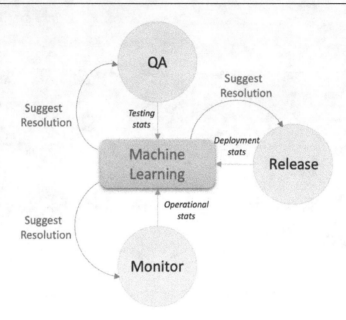

Figure 1.13: Using ML in CI/CD pipeline

Summary

In this chapter, we covered the concepts of MSA and how MSA is different from traditional monolithic architecture. By now, you should also have a clear understanding of the advantages of MSA and the challenges organizations may experience when adopting MSA.

We also covered the key concept of methodologies to consider when designing MSA, such as ACID, the saga pattern, and CQRS. All these concepts are essential to help overcome synchronization challenges and to maintain microservices anonymity.

We now understand the basics of DevOps and why it is important in MSA design, deployment, and operations, as well as how ML integration in MSA enterprise systems can help enhance system operations.

In the next chapter, we will go over common methodologies that organizations pursue to transition from running traditional monolithic systems to MSA systems. We will discuss how to break down the existing system into services that form the new MSA enterprise system.

2

Refactoring Your Monolith

Now we have decided that MSA is the right architectural style for our organization, what's next?

In a recent report, *2022 APIs & Microservices Connectivity Report*, published by Kong Inc., 75% of organizations have a lack of innovation and technology adoption.

The need for an IT system that quickly responds to customer and market needs has never been higher. Monolithic applications can no longer respond to high-paced market updates and needs. That's one main reason for organizations to look to update their IT system, to stay in business.

MSA is a primary enabler for a flexible and reliable enterprise system. Transitioning from a monolithic architecture into MSA is, therefore, becoming essential to modernizing an organization's IT systems.

We will discuss, in this chapter, how to break up the business requirements of an existing running monolithic application in to microservices, and the steps necessary to transition toward MSA applications.

We will cover the following areas as we go over the chapter:

- Identifying the system's microservices
- The ABC monolith
- Function decomposition
- Data decomposition
- Request decomposition

Identifying the system's microservices

Whether it is a brownfield or greenfield enterprise system implementation, we still need to break up business requirements into basic functions as granularly as possible. This will later help us identify each microservice and successfully integrate it into our enterprise system.

In a brownfield system, business and system requirements have already been identified and implemented. They may, however, need to be revisited and updated according to new business criteria, changes, and requirements.

The objective of refactoring your application into simple services is to form highly granular functions that will eventually be built (or acquired) as microservices. You are very likely to add new functions to your new MSA in addition to some of the functions you will already extract from the monolithic system.

We, therefore, split the migration process into the following high-level steps:

1. Define the to-be MSA system and the functions needed to build that MSA.

2. Identify what existing functions in the current monolithic system are to be reused in the new MSA and implemented as microservices.

3. Identify the delta between the existing functions to be reused and the functions needed to get to the to-be MSA system. These are the new functions to be implemented in the new MSA system.

4. From the functions list identified in *step 3*, identify which functions will be developed as a microservice in-house, and the ones that can be acquired through third parties.

Decomposing the monolith using a function-driven approach is a good starting point; nevertheless, using that approach alone is not enough. Since data stores are centralized in the monolith, data dependencies will still be a big concern in maintaining the microservices' autonomy.

The interaction between the different functions in the monolith is another concern. We will need to look into how the function calls are being processed and handled, what data is being shared between these functions, and what data is being returned.

Examining monolithic system functions, data, and function calls (requests) during the refactoring process is essential for maintaining the autonomy of microservices and achieving the desired level of granularity.

> **Important note**
>
> Bear in mind that we must maintain the microservices autonomy principle during the entire monolith decomposition process. Too many microservices would cause a **Nano-service anti-pattern** effect, while too few would still leave your system with the same issues as a monolithic system.

The Nano-service anti-pattern creates too many expectations for most systems' operations, which can in turn further complicate your MSA system and create a lack of stability, decreased reliability, and other system performance issues.

> **Important note**
>
> As a general rule, apply the **Common Closure Principle**, where microservices that change for the same exact reason are better off packaged together in a single microservice.

To better explain the monolith transformation process to an MSA, in the following sections, we will design a simple hypothetical monolithic system, break up the system using the already mentioned three stages of system decomposition, build the different microservices, and then organize them together to build the MSA.

The ABC monolith

ABC is a simplified hypothetical product-ordering monolithic system built specifically to demonstrate the process and the steps needed in refactoring a monolithic application into an MSA. We will be using this ABC system throughout this book to demonstrate some examples of how to apply the concepts and methodologies.

Please note that we put the ABC-Monolith system together for demo purposes only and our aim here is not to discuss how the ABC-Monolith can be designed or structured better. We are more focused on the ABC-Monolith system refactoring process itself.

In the ABC-Monolith, the user can place an order from an existing product catalog and track the order's shipping status. For simplicity, all sales are final, and products cannot be returned.

The system will be able to clear the order payment, assign a shipping courier to the order, and track all order and shipping updates.

The following diagram shows the high-level ABC-Monolith architecture. A user portal is used to add items to the cart, then send the order details to the ABC-Monolith. The ABC-Monolith has different tightly coupled functions with a centralized database, all to process the order from payment to delivery. The user is notified of all order and shipping updates throughout the order fulfillment process.

Figure 2.1: The ABC-Monolith architecture

To further understand the monolith, we will next go over the system As-Is state by discussing the existing monolith's functions, the monolith database structure, and the workflow of the order placement process. We will close this section by comparing the As-Is to the To-Be state.

The ABC-Monolith's current functions

It is imperative to start by understanding what current functions are implemented in the monolith and what their role is in the overall system. The following table lists the system functions we need to consider later in our system refactoring:

Function	Description
place_order()	A function to create a record with all order information, and mark the order as "pending" awaiting the rest of the order placement process.
check_inventory()	To check the availability of an item in the placed order.
process_payment()	Verify the payment of the total order amount. Will return an error code if the payment is not cleared.
update_inventory()	Once an order is verified and the payment is successfully processed, the item inventory should be updated accordingly.
create_order()	The order is now successfully processed; time to change the order status, and kick off the order preparation process (packing, etc.).
create_shipping_request()	Starts the order shipping request and notifies the courier with an available order for shipping.
order_status_update()	A function to update the order status with any changes such as preparing, shipping, exception, received, and so on.
shipment_status_update()	A function to update the shipping status with any changes such as, pending pickup, picked up, en route, exception, received, and so on.
notify_user()	To notify the user of any changes or updates to the placed order.
register_customer()	A function that creates customer record information with a full name, address, phone, and other details.

Table 2.1: The ABC-Monolith functions list

In the preceding table, we focused our description on the role of the function itself regardless of what the parameter passing is, or what the return values are.

The ABC-Monolith's database

All the functions identified in the monolith share a centralized database. The following are the database tables being accessed by the functions:

Database Table	Description
CUSTOMER	A table holding all customer information such as name, email, and phone.
ITEM	The product information is in the catalog. Product information includes product name, price, and stock quantity.
ORDER	Information on orders placed.
ORDER_ITEM	A many-to-many relationship normalization table between the ORDER and ITEM tables.
ORDER_STATUS	The status of each placed order, with a reference to status_code.
STATUS_CODE	Lookup table for order and shipment status codes.
COURIER	Shipping courier information, including courier name, contact, and so on.
SHIPMENT_REQUEST	A list of all shipping requests for orders placed.
SHIPMENT_REQUEST_STATUS	The status of each shipment request, with a reference to status_code.

Table 2.2: The ABC-Monolith database tables list

The following is ABC's **Entity Relationship Diagram** (**ERD**). Note that we needed to create the **ORDER_ITEM** normalization table to break up the many-to-many relationship between both the **ORDER** and **ITEM** tables:

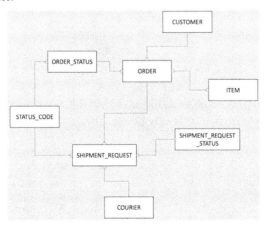

Figure 2.2: The ABC-Monolith ERD

Keep in mind that some of the monolith's functions require full read/write access to specific tables with access to all fields in the table, while some other functions need only access to specific fields in the table. This information is important in system refactoring.

In the following section, we will go over the workflow to identify the ABC-Monolith As-Is state and determine how we can transition into the To-Be state. Along with the workflow information, the function database access requirements will help us refactor the monolith database into individual MSA databases for each microservice.

The ABC workflow and current function calls

We know so far what functions are used in the monolith and how the monolith's database is structured. The next step is to examine the order placement workflow:

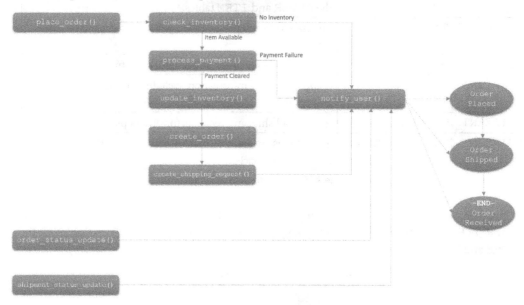

Figure 2.3: The ABC-Monolith function requests/workflow

As shown in the preceding workflow diagram, the individual functions are all executed sequentially. Since it is all one tightly coupled system, there are no synchronization issues expected, and hence no orchestration is needed.

As we move toward the ABC-MSA, however, the decoupling of services creates the need to have a centralized point for managing the execution of these services in a specific sequence.

Shown in the following diagram are ABC's As-Is and To-Be states. No centralized management in the As-Is state is needed; however, an orchestrator component is introduced in the To-Be state to manage the process flows between the services.

Each of the individual services in the To-Be states has a dedicated database, as shown in the diagram. In the As-Is state, on the other hand, the database is centralized.

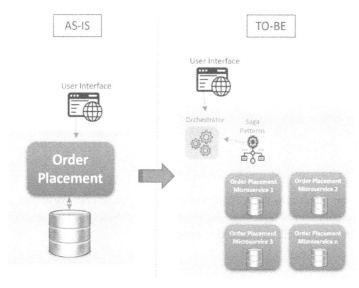

Figure 2.4: The ABC As-Is and To-Be states

Now that we know how our current ABC-Monolith is structured, and what both the as-is and to-be states are, it is time to start the ABC-Monolith refactoring process to transform into the ABC-MSA.

We will refactor the monolith in three stages. First, we will decompose the monolith functions and map these functions to microservices. Then, we will decompose the data to see how the individual databases will be designed. Finally, out of the monolith's workflow, we will analyze the function requests, and build our MSA sagas from there.

Function decomposition

The first step in refactoring the ABC-monolith is to create the microservices based on the system functions we previously identified. This is a straightforward mapping between the existing functions and the microservices.

The key point here is that, by looking only at each function by itself without considering any function calls or data connections, you need to be as granular as possible in your function decomposition.

At first glance, the `notify_user()` function is doing too many things for a microservice, displaying a web user message status/update, notifying the user by email, and/or notifying the user by SMS. Each of these functions can have its own rules, design, issues, and concerns. Splitting the `notify_user()` function into three functions is a better approach from an MSA perspective to achieve the separation of concerns.

Accordingly, we split the `notify_user()` function into one function for handling web messages and notifications, one for handling email notifications, and one for SMS message notifications:

- `web_msg_notification()`
- `email_notification()`
- `sms_notification()`

Similarly, the `process_payment()` function can also be split into two different, more granular functions, one for handling direct credit card payments and one for handling PayPal payments:

- `verify_cc_payment()`
- `verify_paypal_payment()`

The following diagram shows how the ABC-Monolith is broken up so far. We haven't yet looked into how the system's functions are interacting with each other. The function interactions and the order fulfillment's overall workflow will be handled at a later stage.

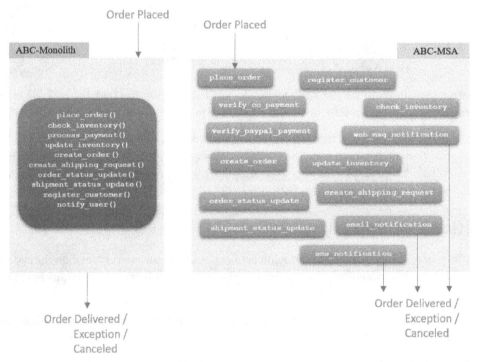

Figure 2.5: The ABC-Monolith function decomposition

At this point, we are satisfied with the current level of granularity so far and we are ready to examine how the database tables are being accessed to see whether further decomposition is needed.

Data decomposition

During this stage, we need to look at how each function is accessing the database and what tables and even which parts of the database tables are accessed.

The following diagram shows what parts of the database the ABC-Monolith functions access. It is essential to know exactly which tables are accessed by which function and why. This will help us identify database dependencies, in order to later eliminate these dependencies and split the centralized ABC-Monolith database into separate data stores, each data store dedicated to each microservice.

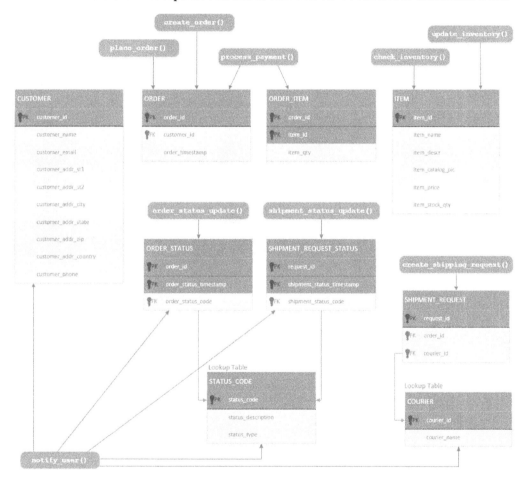

Figure 2.6: ABC-Monolith database access

We are still bound by the microservice autonomy rule. The challenging part in the diagram and this refactoring phase is the shared tables. Sharing a table between two microservices creates coupling that would clearly violate the autonomy rule. On the other hand, creating multiple copies of the table across different microservices will create serious data consistency issues. So, how do we solve this conundrum?

Remember the **saga patterns** that we previously discussed in *Chapter 1*? **Saga patterns** should be able to solve data consistency issues that arise from having a transaction that spans multiple services. In our example here, we can have duplicates of the ORDER table, for example, across the `place_order()`, `create_order()`, and `process_payment()` services of the ABC-MSA system. A similar approach is taken for `check_inventory()`, `update_inventory()`, and so on.

So, with saga patterns in mind, let's reexamine the ABC-Monolith database access shown in the preceding diagram, to build a new database access diagram for services in the ABC-MSA system.

There are two ways to coordinate the data transactions, choreography and orchestration. In choreography, the ABC-MSA saga participant services will have to coordinate data transactions among themselves. In orchestration, a centralized orchestrator performs the coordination process and handles all workflow transactions.

Certainly, we can choose either coordination methodology, but in our example, we would argue that orchestration creates a better decoupling model over choreography. For that reason, and to keep our example simple, we will be using orchestration for our ABC-MSA saga patterns.

The following diagram shows the ABC-MSA service database access. As you can see in the diagram, there are a few database tables that have been copied across the system. We will, in the next section, use saga patterns to maintain data consistency across the copied tables.

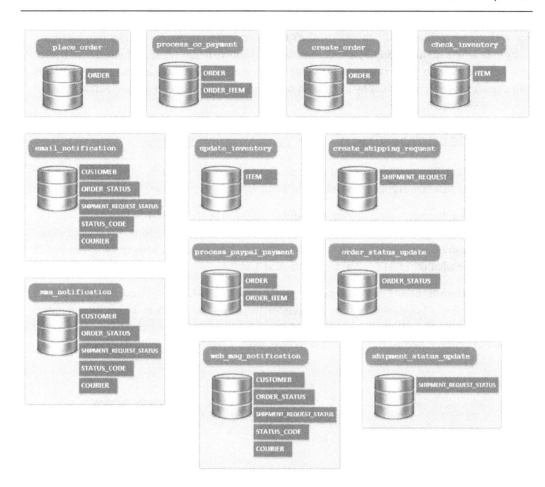

Figure 2.7: ABC-MSA database access

We notice in other services, such as **web_msg_notification**, **email_notification**, and **sms_notification**, that the database is identical for all three services. This is an indication that creating these three services off the original notify_user() function may not be a good idea anyway. You should only see small database access similarities between these different services, not a completely identical database. In a real scenario, we are better off combining these three services into only one service as it originally was.

Similarly, in a real-life scenario, the process_payment() function is likely to be mapped to a single service that includes clearing the payment overall, regardless of whether it is a credit card, PayPal, or any other form of payment. For demo purposes, we will split notify_user() and process_payment() into three and two different services respectively.

So far, we have been able to build the ABC-MSA's microservices from the ABC-Monolith functions, identify data access in the monolith, and decompose the monolith into separate microservices, each with its own database. In the next section, we will focus more on how to ensure isolation and separation of concerns for the microservices by looking into how the service requests are orchestrated in the new ABC-MSA system.

Request decomposition

The ABC-Monolith function request flow has already been identified and shown in *Figure 1.3*. We will now see how this flow is going to work in the ABC-MSA.

In the ABC-MSA, the sagas are programmed and configured in the centralized orchestrator. The orchestrator will initiate separate API calls to each service in the saga, in either a synchronous or asynchronous fashion, depending on the defined workflow, and wait for a response from each API call to determine what other API call(s) to initiate next and how.

The following diagram shows how the workflow would be in the ABC-MSA. Please note that all API calls in our scenario are being initiated from the orchestrator. As you can see from the sequence number, there are some API calls initiated in parallel, and in some other cases, the orchestrator decides the next course of action based on the response it receives from a previously executed service.

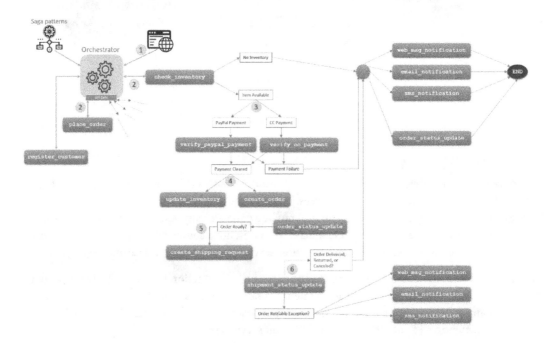

Figure 2.8: The ABC-MSA workflow

The user in the ABC-MSA workflow diagram initiates the order fulfillment process from a web interface, which will kick off the workflow from the orchestrator. Both the **place_order** and **check_inventory** services are launched at the same time by the orchestrator. **place_order** creates the order with all its information and marks its state as **pending**, waiting for the rest of the workflow to be processed.

The **check_inventory** service checks the inventory of items ordered and sends back a **true** or **false** response depending on whether the item is available or not. If any of the items ordered are not available, the **web_msg_notification**, **email_notification**, and **sms_notification** services are triggered.

Now, here is the first challenge: all three notification services will require access to the CUSTOMER database in order to get the customer's name, email address, phone number, and so on. But having one database for all three services creates undesired coupling that would violate the microservices autonomy principle. As we discussed earlier, we should instead create copies of that CUSTOMER database across all services to avoid service coupling. But how do we do that?

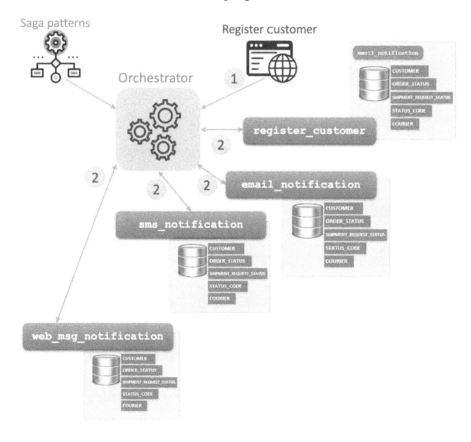

Figure 2.9: Maintaining database consistency across MSA

The CUSTOMER database is mainly managed by the **register_customer** service, which is triggered by the orchestrator through the user interface. To be able to maintain data consistency, and as shown in *Figure 1.9*, the orchestrator will need to simultaneously issue the same transaction on all copies of the CUSTOMER database whenever a record is edited, created, or deleted.

The orchestrator will need to wait for a success confirmation from all four services, **register_customer**, **email_notification**, **sms_ notification**, and **web_ msg_notification**, before the workflow is finalized. Now, what if, let's say, updating the **sms_ notification** CUSTOMER database fails? You will end up with data inconsistency, which can be a serious issue later on.

That's why all saga participants' local transactions will need to have a set of compensating transactions to ensure a rollback in case of any failures in executing the transaction. In our example, the orchestrator will need to undo updates to the CUSTOMER database for all the other services.

The following diagram shows how a failure to update the CUSTOMER database should be rolled back using saga patterns.

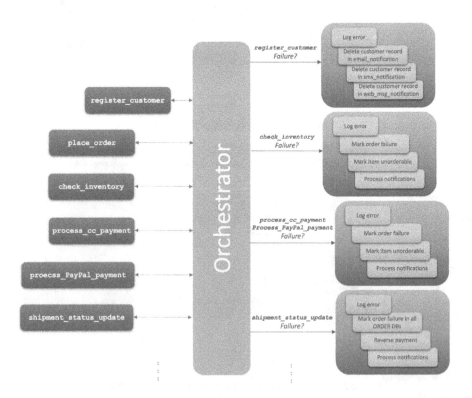

Figure 2.10: Compensating transactions for registering new customer information and placing an order

Summary

In this chapter, we were able to go over the main steps of refactoring a monolith into an MSA, the steps necessary, the main things to consider, and the methodology of doing so. The simplified ABC-Monolith system was a good example; however, as systems get more complicated and the workflow gets more involved, data and process synchronization challenges start to arise.

In *Chapter 1*, we briefly discussed the challenges and the methodologies to be applied to overcome these challenges. In the next chapter, we will start applying the methodology to the ABC system we are trying to refactor.

In the next chapter, we will discuss how we can further maintain microservices' autonomy and MSA stability and overcome some other operational challenges, and the role of API gateways, orchestrators, and microservice aggregators.

3

Solving Common MSA Enterprise System Challenges

In the previous chapter, we learned how to decompose the monolith and refactor it into an MSA enterprise system. We built a simplified system as an example and then refactored the system to demonstrate this process. By doing so, we resolved some of the challenges of running a monolithic system. However, moving toward MSA introduces a completely different set of issues that need to be addressed.

In this chapter, we will discuss the main challenges introduced in MSA, how to address them, and what specific methodologies we need to apply to maintain the MSA system's reliability, durability, and smooth operation.

We will cover the following topics in this chapter:

- MSA system isolation using an Anti-Corruption Layer (ACL)
- API gateways
- Service catalogs and orchestrators
- Microservices aggregators
- Microservices circuit breaker
- Gateways versus orchestrators versus aggregators
- ABC-MSA enhancements

MSA isolation using an ACL

When adopting MSA in a brownfield, your migration from the monolithic system to MSA can either be done as a **big bang migration** or **trickle migration**.

In big bang migration, you keep the old monolith system running as-is while building the entire MSA system. Once the MSA system has been completed, tested, and deployed, you can then completely switch to the new MSA system during your organization's maintenance window, then decommission the old monolith. This type of migration, although usable for some scenarios, is usually not recommended in the case of large enterprise systems.

Switching users from the old to the new system should be done during the corporate's off-peak hours or the corporate's standard migration window. And the sudden switch of users can be a complex and cumbersome process due to the high potential for downtime, potential rollbacks, and risks of unexpected results when applying real traffic to the new system, all of which can impose large time constraints during the migration window.

A common and safer migration approach in our case is trickle migration, where you perform a gradual shift from the old monolithic system to the new MSA system. A common way of doing that is by gradually extracting functions, services, and/or modules out of your monolith and moving them into standalone microservices as part of your new MSA. Gradually, we phase out the existing monolith's functions and build an MSA system piece by piece.

To successfully perform a tickle migration, you need what's called an **Anti-Corruption Layer** (**ACL**), which will act as an intermediate layer, a buffer, and a gateway between the old, messy monolith and your new clean MSA. The ACL layer will help temporarily integrate and glue the new extracted services back into the old system, to be able to communicate with old services, databases, and modules without fouling your new MSA system. You can see the ACL architecture in the following diagram:

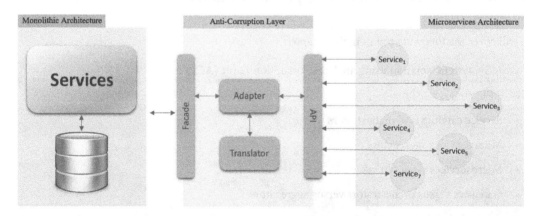

Figure 3.1: Anti-Corruption Layer (ACL)

The ACL lifespan is as long as the monolith system's lifespan. Once the migration has been completed and the monolith has been decommissioned, the ACL will no longer be needed. Therefore, it is recommended that you have the ACL written either as a standalone service or as part of the monolith.

The ACL has three main components:

- The API component, which allows the ACL to communicate with the MSA system using the same language as the MSA system.

- **The ACL Facade**, which is the interface that enables the ACL to talk to the monolith using the monolith's language(s).

    There are two options where the Facade can be placed; one is shown in *Figure 3.1*, where the Facade has been placed as part of the new standalone ACL microservice. The other option is placing the Facade as a component within the monolith, as shown in *Figure 3.2*:

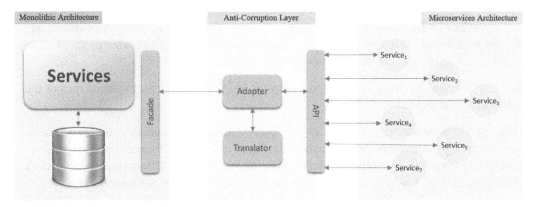

Figure 3.2: The Facade's two implementation options

    The choice would depend on the architects' and developers' preference regarding whether they would like to add more glue code inside the monolith itself, or completely isolate any development effort away from the monolith.

- **The ACL Adapter**, which is a part of the ACL, works between the ACL's northbound API and the Facade. The main function of the Adapter is to translate between the monolith and the MSA using the **ACL Translator** interface, as shown in *Figure 3.1*.

ACL is only needed when a trickle migration is adopted. There is no need for implementing an ACL in the big bang migration case. And since, between both migration styles, there are resources to be consumed, as well as advantages, risks, and tradeoffs, MSA project stakeholders will need to decide on which style is more suitable for the project and the organization.

Whether an ACL is implemented or not, MSA systems would still need a component to act as an interface between the MSA system and external clients. Using an API gateway between the MSA and external API calls is considered a good MSA design practice. The next section discusses the roles of the API gateway in MSA systems, and the tradeoffs of having to adopt an API gateway in MSA design.

Using an API gateway

As we explained in *Chapter 1*, microservices can communicate directly with each other without the need for a centralized manager. As the MSA system becomes more mature, the number of microservices gradually increases, and direct communication between microservices can become a large overhead – especially with calls that need multiple round trips between the API consumer and the API provider.

With the microservices' autonomy principle, each microservice can use its technology stack and may communicate with a different API contract than the other microservices in the same MSA system. One microservice, for example, may only understand a RESTful API with a JSON data structure, while others may only communicate with Thrift or Avro.

Moreover, the location (IP and listening port) of the active instantiated microservices change dynamically within the MSA system. Therefore, the system will need to have a mechanism to identify the location at which the API consumer can point its calls towards.

There are also situations where you need to tie in your MSA system to legacy systems such as the mainframe, AS400, and more.

All of the previous situations require code to be embedded in each microservice in the MSA system. This code will help the microservices understand the legacy and non-REST communication patterns, discover the network location of other microservices in the system, and understand each microservice's needs in general. Now, how independent and portable would such a set of microservices be?

A better approach to addressing the preceding challenges is to use an API gateway where all system services talk to each other through that gateway. The API gateway receives API calls from the system's API consumers, then maps the data received into a data structure and a protocol that API providers can understand and process:

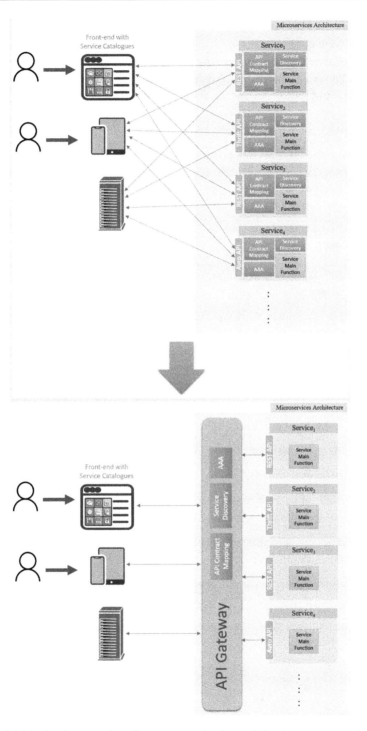

Figure 3.3: Moving from services direct communication to API gateway communication

With the API gateway, we significantly reduce direct 1-to-1 communication between services. Moreover, we offload the system's microservices from having multiple translations, mapping code, and **Authentication-Authorization-Accounting (AAA)** tasks. Rather, we move the responsibility of discovering the location of microservices from the client to the API gateway, which, in turn, further reduces the code overhead and renders microservices as light and independent as possible.

> **Important note**
>
> The MSA system's availability is as good as the API gateway's availability. Therefore, it is necessary for the API gateway to be developed, deployed, and managed as a high-performance and highly available mission-critical service.

The API gateway can be deployed as a standalone service that's part of the MSA system. The main functions of the gateway are as follows:

- Minimize the API calls between microservices, which makes MSA inter-service communication much more efficient.

- Minimize API dependencies and breaking changes. In an MSA system with no API gateway, if, for whatever reason, one of the API providers changes its API, a breaking change will likely happen. This means we will need to create a change in every microservice communicating with the API provider. By using an API gateway, a change in the API provider will be limited to the API gateway only, to match the API contract between the provider and the consumers.

- Translate and map between API contracts, which offloads the microservices from embedding translation code in their core function.

- Run a service discovery mechanism and offload clients from running that function.

- Act as the entry point to the MSA system's external client calls.

- Load-balance API calls across the different instances of high-availability microservices, and offload microservices in high-traffic situations.

- Offer better security by throttling sudden increases in API calls during **Distributed Denial of Service (DDoS)** and similar attacks.

- Authenticate and authorize users to access different components in the MSA system.

- Provide comprehensive analytics to provide deep insights into system metrics and logs, which can help further enhance system design and performance.

Despite all these advantages and functions of the API gateway, there are still some drawbacks to having an API gateway in your MSA system:

- The most obvious is complexity. The more protocols and API contract data structures we have in the system, the more complex the gateway becomes.

- The MSA's operation is highly dependent on the API gateway's performance and availability, which may create an unwanted system performance bottleneck.

- Introducing an additional intermediary component such as an API gateway in the path of intra-microservice communication increases service response time. And with chatty services, the increased response time can become considerable.

Even with all the functions the API gateway provides, we still need a way to map each user request to specific tasks that the MSA system would need to run to fulfill that request. In the next section, we'll discuss how the MSA system tasks are mapped to specific user services, and how these tasks are orchestrated in MSA.

Service catalogs and orchestrators

Orchestration is one of the most commonly used communication patterns in MSA systems. We briefly discussed this concept in *Chapter 1*. In this section, we will dive into more details about orchestration.

Determining the most appropriate communication pattern between the different microservices depends on many factors. Among the factors that will help determine whether choreography or orchestration is the most suited communication pattern for the system, you must consider the number of microservices you have in the system, the level of interactions between the different microservices, the business logic itself, how dynamic business requirements change, and how dynamic system updates are.

Orchestrators act as the central managers controlling all communication between the system's microservices. They usually interact with the users through a dashboard interface that contains all service catalogs. The **Service Catalog** is a set of services the MSA system offers to users. Each service in the catalog is linked to a set of workflows. **Workflows** are the actions the orchestrator will trigger and coordinate between the system's microservices to deliver the service the user selected from the catalog:

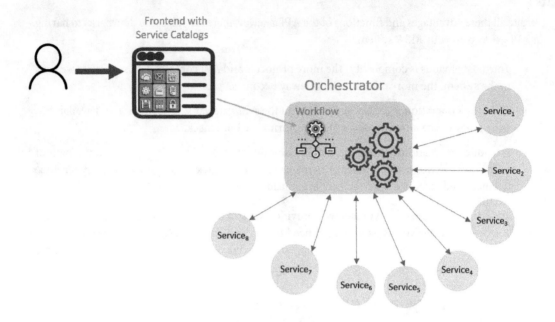

Figure 3.4: Orchestrators in MSA

The orchestrator's functions can be extended beyond managing workflows. Orchestrators can also manage the entire life cycle of microservices; this involves provisioning and deploying microservices, configuring the microservice, and performing upgrades, updates, monitoring, performance audits, and shutdowns when needed.

> **Important note**
>
> The orchestrator is the main brain of the MSA system, and it is imperative to have the orchestrator deployed and managed as a high-performing and mission-critical component of the MSA system.

Some of the benefits of running an orchestrator in MSA include the following:

- You have a centralized management platform as a single source of truth for all of your workflows. Thus, you can build complex workflows in a complex MSA without having to worry about how many microservices you have and how they can scale.

- As in the API gateway, you can tie in your legacy systems or part of your old monolith and completely isolate your microservices from having to couple with any other system component. This saves a lot of effort having to build code into the independent microservices and tremendously helps in scaling your MSA.

- Microservices are visible to the orchestrator and hence can be completely managed, audited, and monitored by the orchestrator. This can produce very helpful and insightful analytics that can further enhance the MSA system's supportability and operations.

- The orchestrator's visibility can help in troubleshooting any operational issues and identifying problems quickly.

- Orchestrators can automatically detect and self-resolute some of the operational problems. Orchestrators can, for example, detect resource starvation and reroute requests to a backup microservice. Orchestrators can automatically vertically or horizontally scale a particular microservice when a problem is detected. Orchestrators can also try to automatically restart the service if the service is not responding.

The orchestrator solves many of the MSA operational problems, including some of the data synchronization challenges. When scaling the system, however, data synchronization and data consistency become a big challenge for the orchestrator to address by itself. Microservices aggregators help address data synchronization issues when the MSA system scales. In the next section, we will discuss what the aggregator pattern is, what it is used for, and how it works.

Microservices aggregators

In *Chapter 2*, we had to copy some schemas across multiple microservices and use saga patterns through the orchestrator to keep data consistent and preserve the microservice's autonomy. This solution may be viable in a situation if you have a limited number of microservices within the MSA system. In a large number of microservices systems, copying schemas across different microservices to maintain the synchronization of the individual microservices database doesn't scale well and can severely impact the system's overall performance.

Consider an MSA system with 100 microservices and copy schemas across about 20 of those microservices to maintain the microservice's autonomy. Each time any part of any of the schema's data is updated, the orchestrator will have to sync those 20 schemas.

Moreover, even if we have all 100 microservices perfectly autonomous, what if one of the user's operations needs to gather information from those 20 microservices? The orchestrator will have to issue at least 20 different API calls to 20 different microservices to get the information the user is looking for. Not to mention that some of these 20 microservices may need to exchange multiple API calls to send the result back to the user.

To put things into perspective, let's revisit the ABC-MSA system we built in *Chapter 2*. We have the `order`, `product`, and `inventory` microservices. The `product` microservice is for managing product information, the `inventory` microservice is for managing the product inventory, and the `order` microservice is for placing and managing orders.

Let's assume we're in a situation where a sales analyst is generating a report to check a product's average quantity purchased per order and the product's inventory level at the time at which the order was placed, as shown in the following diagram:

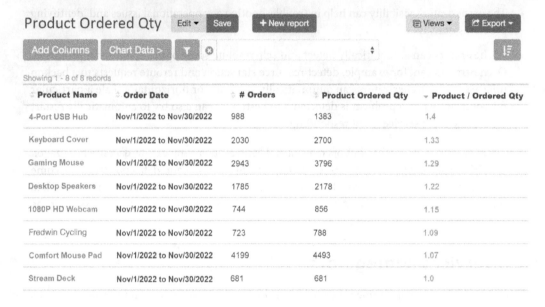

Product Name	Order Date	# Orders	Product Ordered Qty	Product / Ordered Qty
4-Port USB Hub	Nov/1/2022 to Nov/30/2022	988	1383	1.4
Keyboard Cover	Nov/1/2022 to Nov/30/2022	2030	2700	1.33
Gaming Mouse	Nov/1/2022 to Nov/30/2022	2943	3796	1.29
Desktop Speakers	Nov/1/2022 to Nov/30/2022	1785	2178	1.22
1080P HD Webcam	Nov/1/2022 to Nov/30/2022	744	856	1.15
Fredwin Cycling	Nov/1/2022 to Nov/30/2022	723	788	1.09
Comfort Mouse Pad	Nov/1/2022 to Nov/30/2022	4199	4493	1.07
Stream Deck	Nov/1/2022 to Nov/30/2022	681	681	1.0

Figure 3.5: A sample product order report

The orchestrator will have to send at least three API calls, one to each of the `order`, `product`, and `inventory` microservices, as shown in the following diagram:

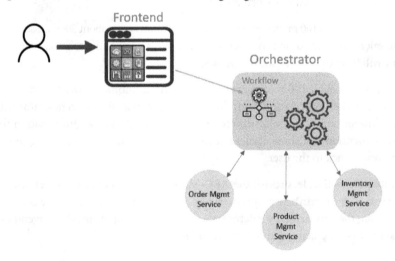

Figure 3.6: A user operation spanning multiple microservices

To minimize dependencies and response time, a better approach for this particular situation is to use an **aggregator**. This aggregator will collect the different pieces of data from all `order`, `product`, and `inventory` microservices and update its database with the combined information.

The API gateway or consumer will only need to send one API call to the aggregator to get all the information it needs. The number of API calls is minimized and the overall response time is greatly reduced, especially in cases where the information required is distributed across a large number of microservices:

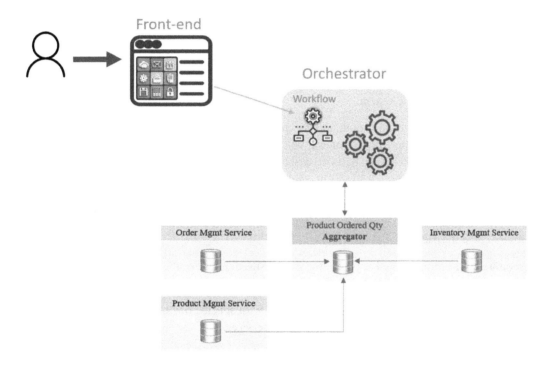

Figure 3.7: An aggregator communication pattern

The aggregator communication pattern reduces the number of API calls users could trigger in various operational requests and further enhances the data synchronization's design and performance, as well as the overall system performance, especially in high-latency networks.

Now, we know the roles of the API gateway, the orchestrator, and the aggregator. In the next section, we will discuss how all these three components interact with each other in the MSA.

Gateways versus orchestrators versus aggregators

From what we have described so far, there are some overlapping functions between the API gateway, the orchestrator, and the aggregator. In this section, we will answer some of the fundamental questions regarding how all three components interact in a single MSA system:

- How do these three MSA components work together?
- Can an API gateway perform the aggregator and orchestrator functions?
- What are the best practices for deploying all these communication patterns in our MSA?

First of all, theoretically speaking, you can have clients interact with the MSA microservices directly without an API gateway. However, this would not be a good practice. By having no gateway in your MSA system, you would need to have most of the gateway functions implemented within each microservice you have in the system.

To keep microservices as light and autonomous as possible, it is highly recommended to have an API gateway in your MSA system. The API gateway will handle all ingress and egress API traffic from the different types of clients. Clients can be a web dashboard, a mobile application, a tablet, a third-party integration system, and so on.

Whether you add an aggregator or not will highly depend on your business logic and system design. You will only need aggregators if you have client use cases where requests need to span across multiple microservices at the backend.

Aggregators can be implemented as part of the gateway itself; however, the best practice is to add an aggregator only whenever it is needed and make it an independent standalone microservice.

An MSA system can have multiple aggregators, each with specific business logic, and fulfilling a specific collection of data from a different set of microservices.

Similarly, orchestration patterns can also be implemented in the API gateway; however, you need the API gateway to focus on doing the main functions it was created for and leave orchestration tasks to the orchestrator. The orchestrator is also best deployed as a standalone microservice:

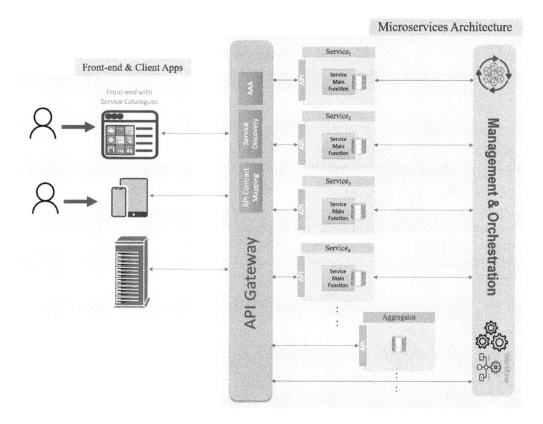

Figure 3.8: MSA high-level architecture

The preceding diagram shows the high-level architecture of having all these components working together within the MSA system. Clients always interact with the API gateway, and the API gateway will route the request to the appropriate service within the MSA system.

Client API calls are routed to the appropriate microservice based on the API's configuration. If the client request is something that is fulfilled by communicating with a single microservice, then the gateway will send that request directly to the microservice. API calls that span multiple microservices and are assigned to a specific aggregator in the system will be forwarded to that particular aggregator. Finally, for API requests that invoke specific workflows, the API gateway will forward those to the orchestrator.

In this section, we learned how the API gateway, the orchestrator, and the aggregator coexist in the same MSA system. In the next section, we will try to apply the concepts of all these three communication patterns to our previously developed ABC-MSA system.

Microservices circuit breaker

Another challenge in MSA systems is the stability and assurance of workflow execution. Saga patterns, which we discussed in *Chapter 1*, are used to ensure that all transactions within a specific workflow are either all successfully executed, or all fail. But is that enough to ensure the reliable execution of microservices?

Let's consider a scenario where the called microservice is too slow in responding to API calls. Requests get successfully executed, but the microservice's response times out. The microservice consumer, in turn, may assume an execution failure, and accordingly repeat the operation, which can be very problematic.

As shown in the following diagram, when a response timeout takes place in the Payment microservice, the Payment microservice will process the payment, but the microservice consumer will assume that the payment has not been processed and may automatically (or upon user request) retry the process. This behavior will cause the payment to be processed multiple times, resulting in multiple charges for the same order, or for the order to be placed multiple times:

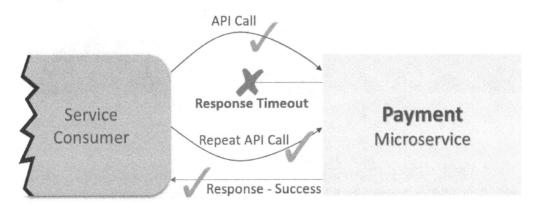

Figure 3.9: Payment microservice with too slow of a response time

In MSA, when microservices get instantiated, they start with limited resources and threads to avoid one particular microservice from hogging all the system's resources.

With system resources in mind, consider another scenario, as shown in *Figure 3.10*, where the Inventory microservice is part of a service workflow and the Payment microservice is neither processing nor responding to API calls for whatever reason. In this case, both the Order and Payment microservices will keep waiting for confirmation from the inventory before they start releasing their resources.

With the Inventory microservice timing out requests, and under a system heavy load or high order volume, requests start to pile up for the Order and Payment microservices. Eventually, both the Order and Payment microservices start to run out of resources and become unable to respond to requests:

Figure 3.10: Inventory microservice is down

Similar scenarios in MSA can result in a domino effect, causing a cascading failure to multiple microservices, which, in turn, causes an entire system failure.

A microservice **circuit breaker** is used to prevent a system cascading failure from happening. A circuit breaker monitors microservice performance using real traffic metrics. It analyses parameters such as response time and successful response rate and then determines the health of the microservice in real time. Should the microservice become unhealthy, the circuit breaker immediately starts responding to the microservice consumers with an error.

A circuit breaker does not prevent the microservice being monitored from failing; rather, it averts a cascading failure from taking place:

Figure 3.11: The Inventory microservice with an inline circuit breaker

When the circuit breaker assumes a microservice is unhealthy, it still needs to monitor and evaluate the microservice's operational performance. The circuit breaker switches to a **half-open state**, where it only allows a small portion of requests to pass through to the microservice being monitored. Once the circuit breaker detects a healthy microservice response, the circuit breaker switches its state back to a **closed state**, where API traffic flows back normally to the microservice.

Circuit breakers are not needed for every single microservice in the MSA system. We only need to deploy a circuit breaker on microservices that can cause a cascading failure. Architects will need to study and determine which microservices need circuit breaker protection.

A circuit breaker can be deployed as a standalone microservice or be part of the API gateway. Whether circuit breakers are implemented as standalone microservices or part of the API gateway highly depends on the system's business and operational requirements, as well as the patterns adopted in the architecture itself.

The circuit breaker pattern, as well as the orchestrator, aggregator, ACL, and API gateway, are all enhancements architects that can be applied to the MSA system for better reliability, resilience, and overall performance. In the next section, we will learn how to apply each of the patterns discussed here to our ABC-MSA system.

ABC-MSA enhancements

In *Chapter 2*, we refactored our ABC-Monolith into a simple ABC-MSA. The ABC-MSA we designed in *Chapter 2* lacked many of the enhancements we are considering in this chapter. It is time to take what we have learned in this chapter and apply that to the ABC-MSA system to enhance its design and operations.

First of all, in the ABC-MSA from *Chapter 2*, the orchestrator was doing both the API gateway functions and the orchestration function. So far, we have learned that combining both the gateway and orchestration functions in one service is not the best option. Therefore, we will add to our ABC-MSA system an API gateway dedicated to ingress and egress API calls, and other API gateway functions we discussed earlier in this chapter, such as authentication, authorization, audit, monitoring, and so on.

The API gateway will run as a separate standalone microservice serving direct client requests, including the system dashboard and user frontend. The orchestrator will also run as a standalone microservice serving the MSA's workflows.

The aggregator(s) will depend on the use cases where multiple ABC-MSA microservices are used to fulfill the user requests.

A simple use case for using an aggregator would be a user checking for an order's shipping status. The status should include the order information, the products included in the order, and the shipping status of that order.

To show all this information to the user, we will need to pull information from three different microservices: Order Management, Product Management, and Shipping Management. We will deploy an aggregator as a standalone microservice to pull the data from all these microservices and make them available for user API consumption:

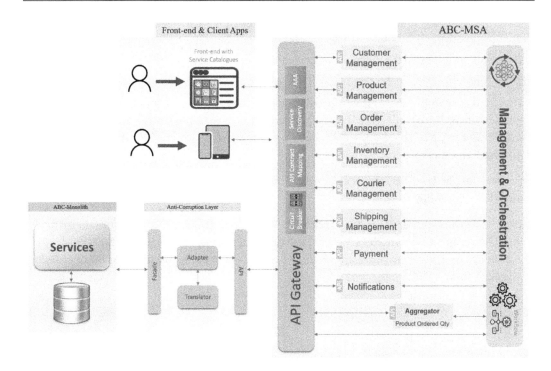

Figure 3.12: The enhanced ABC-MSA architecture

In the preceding diagram, we added the Management and Orchestration layer as part of the ABC-MSA system. This layer will manage the orchestration workflows and the microservice's life cycle, including, installation, configuration, instantiation, updates, upgrades, and shutdowns.

We will also need an ACL to be active during the transition from the ABC-Monolith to the ABC-MSA. The ACL will act as a buffer between both systems to maintain the neatness of the architecture and its operations. Once all the ABC-Monolith functions have been redeployed into the ABC-MSA, both the old ABC-Monolith and the ACL can be decommissioned.

Summary

In this chapter, we discussed the different components of the MSA that can be introduced to maintain the system's stability and enhance its performance.

We discussed how to use the ACL to protect our new MSA during its transition from the old monolithic system. Then, we covered the roles and functions of the API gateway, the aggregator, and the orchestrator. We also covered some of the drawbacks you may experience when adopting the various communication patterns in MSA.

Finally, we redesigned our ABC-MSA to showcase how these different components can all function together in a typical MSA.

Chapter 1 to *Chapter 3* covered the basics of the MSA. In the next chapter, we will start discussing, with hands-on examples, key machine learning and deep learning algorithms used in MSA enterprise systems, and go over some programming and tool examples of building machine learning and deep learning algorithms.

Part 2: Overview of Machine Learning Algorithms and Applications

In this part, we will shift our focus to machine learning. We will learn about the different concepts of machine learning algorithms and how to design and build a machine learning system, maintain the model, and apply machine learning to an intelligent enterprise MSA.

We will first learn the fundamentals when it comes to identifying the difference between different machine learning models and their use cases. Once we've covered the basics, we will start to learn how to design a machine learning system pipeline. Once we have established a machine learning system pipeline, we will learn what data shifts are, how they can impact our system, and how we can identify and address them. Finally, having gone through all the basics, we will start to explore the different use cases for building our very own intelligent enterprise MSA.

By the end of *Part 2*, we will have a basic understanding of machine learning and the different algorithms, how to build and maintain a machine learning system, and, finally, the different use cases in which we can use machine learning for our intelligent enterprise MSA.

This part comprises the following chapters:

- *Chapter 4, Key Machine Learning Algorithms and Concepts*
- *Chapter 5, Machine Learning System Design*
- *Chapter 6, Stabilizing the Machine Learning System*
- *Chapter 7, How Machine Learning and Deep Learning Help in MSA Enterprise Systems*

4

Key Machine Learning Algorithms and Concepts

In the previous chapters, we explored the different concepts of MSA and the role it plays when creating enterprise systems.

In the coming chapters, we will begin to shift our focus from learning about MSA concepts to learning about key machine learning concepts. We will also learn about the different libraries and packages being used in machine learning models using Python.

We will cover the following areas in this chapter:

- The differences between **artificial intelligence**, **machine learning**, and **deep learning**
- Common deep learning packages and libraries used in Python
- Building **regression** models
- Building multiclass **classification**
- Text sentiment analysis and topic modeling
- Pattern analysis and forecasting using machine learning
- Building enhanced models using deep learning

The differences between artificial intelligence, machine learning, and deep learning

Despite the recent rise in popularity of artificial intelligence and machine learning, the field of artificial intelligence has been around since the 1960s. With different sub-fields emerging, it is important to be able to differentiate between them and understand them and what they entail.

To start, artificial intelligence is the overarching field that encompasses all the sub-fields we see today, such as machine learning, deep learning, and more. Any system that perceives or receives information from its environment and carries out an action to maximize the reward or achieve its goal is considered to be an artificially intelligent machine.

This is commonly used today when it comes to robotics. Most of our machines are designed so that they can capture data using their sensors, such as cameras, sonars, or gyroscopes, and use the data captured to respond to a particular task most efficiently. This concept is very similar to how humans function. We use our senses to "capture" information from our environment and based on the information we receive, we carry out certain actions.

Artificial intelligence is an expansive field, but it can be broken into different sub-fields, one we commonly know today as machine learning. What makes machine learning unique is that this field works on creating systems or machines that can continually learn and improve their model without explicitly being programmed.

Machine learning does this by collecting data, also known as training data, and trying to find patterns in the data to make accurate predictions without being programmed to do so. There are many different methods used in machine learning to learn the data and the methods are tailored to the different problems we encounter.

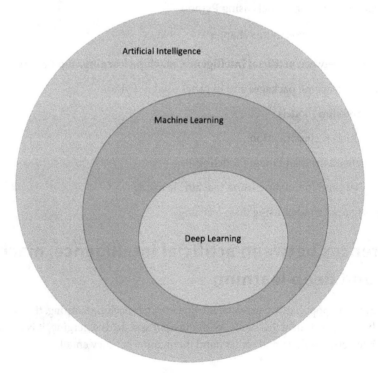

Figure 4.1: Different fields in artificial intelligence

Machine learning problems can be broken down into three different tasks: **supervised learning**, **unsupervised learning**, and **reinforcement learning**. For now, we will focus on supervised and unsupervised learning. This distinction is based on the training data that we have. Supervised learning is when we have the input data and the expected output for the particular set of data, which is also called the label. Unsupervised learning, on the other hand, only consists of the input without an expected output.

Supervised learning works by understanding the relationship between the input and output data. One common example of supervised learning is predicting the price of a home in a certain city. We can collect data on existing homes by capturing their specifications and their current prices and then learn the pattern between the characteristics of these homes and their prices. We can then take a home, not in our training set, and test our model by inputting the features of the house into our program and have the model predict the price of the home.

Unsupervised learning works by learning about the structure of the data either using grouping or clustering methods. This method is commonly used for marketing purposes. For example, a store wants to cluster its customers into different groups so that it can efficiently tailor its products to different demographics. It can capture the purchase history of its customers, use that data to learn about purchasing patterns, and suggest certain items or goods that would interest them, thus maximizing its revenue.

Before we can understand deep learning, which is a sub-field of machine learning, we must first understand what **Artificial Neural Networks** (**ANNs**) are. Taking inspiration from neurons in a brain, ANNs are models that comprise a network of fully connected nodes, also known as artificial neurons. They contain a set of inputs, hidden layers connecting the neurons, and also an output node. Each neuron has an input and output, which can be propagated throughout the network. In order to calculate the output of a neuron, we take the weighted sum of all the inputs, multiply it by the weight of the neuron, and then usually add a bias term.

We continue to perform these actions until we reach the last layer, which is the output neuron. We perform a nonlinear activation function, such as a sigmoid function, to give us the final prediction. We then take the predicted output value and input it in a **cost function**. This function tells us how well our network is learning. We take this value and backpropagate through our layers back to the first layer, adjusting the weights of the neurons depending on how our network is performing. With this, we can create strong models that can perform tasks such as handwriting recognition, game-playing AI, and much more.

> **Important Note**
>
> A program is considered to be a machine learning model if it can take input data and learn the patterns to make predictions without being explicitly programmed to.

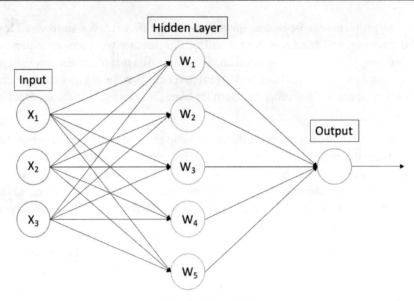

Figure 4.2: An ANN

While ANNs are capable of performing many tasks, there are significant downsides that limit their use in today's market:

- It can be difficult to understand how the model performs. As you add more hidden layers to the network, it becomes complicated to try and debug the network.

- Training the model takes a long time, especially with copious amounts of training data, and can drain hardware resources, as it is difficult to perform all these mathematical operations on a CPU.

- The biggest issue with ANNs is overfitting. As we add more hidden layers, there is a point at which the weights assigned to the neurons will be heavily tailored to our training data. This makes our network perform very poorly when we try to test it with data it has not seen before.

This is where deep learning comes into play. Deep learning can be categorized by these key features:

- **The hierarchical composition of layers**: Rather than having only fully connected layers in a network, we can create and combine multiple different layers, consisting of non-linear and linear transformations. These different layers play a role in extracting key features in the data that would be otherwise difficult to find in an ANN.

- **End-to-end learning**: The network starts with a method called feature extraction. It looks at the data and finds a way to group redundant information and identify the important features of the data. The network then uses these features to train and predict or classify using fully connected layers.

- **A distributed representation of neurons**: With feature extraction, the network can group neurons to encode a bigger feature of the data. Unlike in an ANN, no single neuron encodes everything. This allows the model to reduce the number of parameters it has to learn while still retaining the key elements in the data.

Deep learning is prevalent in computer vision. Due to the advances in the technology of capturing photos and videos, it has become very difficult for ANNs to learn and perform well when it comes to image detection. For starters, when we use an image to train our model, we have to look at every pixel in an image as an input to the model. So, for an image of resolution 256x256, we would be looking at over 65,000 input parameters. Depending on the number of neurons in your fully connected layer, you could be looking at millions of parameters. With the sheer number of parameters, this will be bound to cause overfitting and could take days of training.

With deep learning, we can create a group of layers called **Convolutional Neural Networks** (**CNNs**). These layers are responsible for reducing the number of parameters that we have to learn in our model while still retaining the key features in our data. With these additions, we can learn how to extract certain features and use those to train our model to predict with efficiency and accuracy.

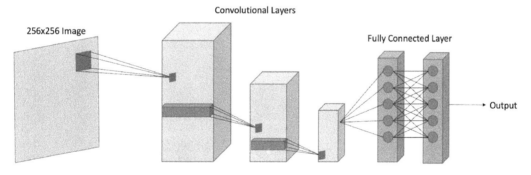

Figure 4.3: A CNN

In the next section, we will be looking at the different Python libraries used for machine learning and deep learning and their different use cases.

Common deep learning and machine learning libraries used in Python

Now that we have gone over the concepts of artificial intelligence and machine learning, we can start looking at the programming aspect of implementing these concepts. Many programming languages are used today when it comes to creating machine learning models. Commonly used are MATLAB, R, and Python. Among them, Python has grown to be the most popular programming language in machine learning due to its versatility as a programming language and the extensive number of libraries, which makes creating machine learning models easier. In this section, we will be going over the most commonly used libraries today.

NumPy

NumPy is an essential package when it comes to building machine learning models in Python. You will be mostly working with large, multi-dimensional matrices when building your models. Most of the effort is spent on transforming, splicing, and performing advanced mathematical operations on matrices, and NumPy provides the tools need to perform these actions while retaining speed and efficiency.

For more information on the different APIs that NumPy offers, you can visit the documentation on its website: `https://numpy.org/doc/stable/reference/index.html`.

Here, we will look at the example code. This section shows us how we can initialize a NumPy array. In this example, we will create a 3x3 matrix with initialized values of 1 through 9:

```python
import numpy as np
# creates a 3x3 numpy array
arr = np.array([[1,2,3],[4,5,6],[7, 8, 9]])
```

Here, we will print out the results:

```python
print(arr)
[[1 2 3]
 [4 5 6]
 [7 8 9]]
```

Now, we can show how we can splice and extract certain elements from our array.

This line of code allows us to pull all the values that are in the second column of our array. Keep in mind that in NumPy our arrays and lists are zero-indexed, meaning that the zero index refers to the first element in the array or list:

```python
print(arr[:,1]) # print the second column of the array
[2 5 8]
```

In this example, we extract all of the values in the row of 2 in our array:

```python
print(arr[2,:]) # print the last row of the array
[7 8 9]
```

Another useful aspect of NumPy arrays is that we can apply mathematical functions to our matrices without having to implement code to perform basic functions. Not only is this much easier but it also is much faster and more efficient.

In this example, we simply perform a multiplication between our matrix and a scalar value of -1:

```
print(np.multiply(arr, -1)) # multiplies every element in the
array by -1
[[-1 -2 -3]
 [-4 -5 -6]
 [-7 -8 -9]]
```

Matplotlib

In order to see how your model is learning and performing, it is important to be able to visualize your results and your data. Matplotlib offers a simple way to graph your data, from something as simple as a line plot to more advanced plots, such as contour plots and 3D plots. What makes this library so popular is its seamlessness when working with NumPy.

For more information on their different functions, you can visit their website: `https://matplotlib. org/stable/index.html`.

In this example, we will create a simple line graph. We first initialize two arrays, x and y, and both arrays will contain values from 0 to 9. Then, using Matplotlib's APIs, we can plot and show our simple graph:

```
import matplotlib.pyplot as plt
import numpy as np
x = np.arange(10) # creates an array from 0-9
y = np.arange(10)
plt.plot(x,y)
plt.show()
```

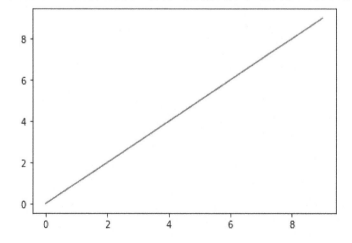

Figure 4.4: A simple line graph using Matplotlib

Pandas

With the recent trend of storing data in CSV files, Pandas has become a staple in the Python community due to its ease and versatility. Pandas is commonly used for data analysis. It stores the data in a tabular format, and it provides users with simple functions to pre-process and manipulate the data to fit their needs. It has also become useful when dealing with time-series data, which is helpful when building forecasting models.

For more information on the different functions, you can view the documentation on its website: `https://pandas.pydata.org/docs/`.

In this example, first, we will simply initialize a DataFrame. This is the data structure used to store our data in a two-dimensional tabular format in Pandas. Usually, we store the data from the files we read from, but it is also possible to create a DataFrame with your own data:

```python
import pandas as pd
data = {
    "Number of Bedrooms": [5, 4, 2, 3],
    "Year Build": [2019, 2017, 2010, 2015],
    "Size(Sq ft.)": [14560, 12487, 9882, 10110],
    "Has Garage": ["Yes", "Yes", "No", "Yes"],
    "Price": [305000, 275600, 175000, 235000],
}
df = pd.DataFrame(data)
print(df)
```

```
   Number of Bedrooms  Year Build  Size(Sq ft.) Has Garage   Price
0                   5        2019         14560        Yes  305000
1                   4        2017         12487        Yes  275600
2                   2        2010          9882         No  175000
3                   3        2015         10110        Yes  235000
```

Figure 4.5: Output of our DataFrame

As with NumPy, we can extract certain columns and rows of our DataFrame. In this code, we can view the first row of our DataFrame:

```python
print(df.iloc[0]) # view the first entry in the table
```

```
Number of Bedrooms           5
Year Build                2019
Size(Sq ft.)             14560
Has Garage                 Yes
Price                   305000
Name: 0, dtype: object
```

Figure 4.6: Output of the first row of our DataFrame

With Pandas, we can also extract certain columns from our DataFrame by using the name of the column rather than the index:

```
print(df["Price"]) # print all the values in the Prices column
```

```
0     305000
1     275600
2     175000
3     235000
Name: Price, dtype: int64
```

Figure 4.7: Output of all the values in the Price column

TensorFlow and Keras

TensorFlow and Keras are the foundation when it comes to building deep learning models. While both can be used individually, Keras is used as an interface for the TensorFlow framework, allowing users to easily create powerful deep learning models.

TensorFlow, created by Google, functions as the backend when creating machine learning models. It works by creating static data flow graphs that specify how the data moves through the deep learning pipeline. The graph contains nodes and edges, where the nodes represent mathematical operations. It passes this data using multidimensional arrays known as Tensors.

Keras, later to be integrated with TensorFlow, can be viewed as the frontend for designing deep learning models. It was implemented to be user-friendly by allowing users to focus on designing their neural network models without having to deal with a complicated backend. It is similar to object-oriented programming, as it replicates the style of creating objects. Users can freely add different types of layers, activation functions, and more. They can even use prebuilt neural networks for easy training and testing.

In the following example code, we can see how we can create a simple, hidden two-layer neural network. This block of code allows us to initialize a `Sequential` model, which consists of a simple stack of layers:

```
import tensorflow as tf
from tensorflow import keras
```

```
from tensorflow.keras import layers
from keras.models import Sequential
model = Sequential()
model.add(Flatten(input_shape=[256,256]))
```

Depending on our application, we can add multiple layers with different configurations, such as the number of nodes, the activation functions, and the kernel regularizer:

```
 #Adding First Hidden Layer
model.add(tf.keras.layers.Dense(units=6,kernel_
regularizer='l2',activation="leaky_relu"))
 #Adding Second Hidden Layer
model.add(tf.keras.layers.Dense(units=1,kernel_
regularizer='l2',activation="leaky_relu"))
#Adding Output Layer
model.add(tf.keras.layers.Dense(units=1,kernel_
regularizer='l2',activation="sigmoid"))
```

Finally, we can compile our model, which essentially gathers all the different layers and combines them into one simple neural network:

```
#Compiling ANN
model.compile(optimizer='sgd',loss="binary_
crossentropy",metrics=['accuracy'])
```

PyTorch

PyTorch is another machine learning framework created by Meta, formally known as Facebook. Much like Keras/TensorFlow, it allows the users to create machine learning models. The framework is well suited to **Natural Language Processing** (**NLP**) and computer vision problems but can be tailored to most applications. What makes PyTorch unique is its dynamic computational graph. It has a module called Autograd, which allows you to perform automatic differentiation dynamically, compared to TensorFlow, in which it is static. Also, PyTorch is more in line with the Python language, which makes it easier to understand and takes advantage of useful features of Python such as parallel programming. For more information, visit the documentation on their website: https://pytorch.org/docs/stable/index.html.

In this section of code, we can create a simple single-layer neural network. Similar to Keras, we can initialize a Sequential model and add layers depending on our needs:

```
import torch
model = torch.nn.Sequential( # create a single layer Neural
```

```
Network
    torch.nn.Linear(3, 1),
    torch.nn.Flatten(0, 1)
)
loss = torch.nn.MSELoss(reduction='sum')
```

SciPy

This library is designed for scientific computing. There are many built-in functions and methods used for linear algebra, optimization, and integration, which are commonly used in machine learning. This library is useful when trying to compute certain statistics and transformations as you build your machine learning model. For more information on the different functions it provides, view the documentation on its website: https://docs.scipy.org/doc/scipy/.

In this example code, we can create a 3x3 array using NumPy and then we can use SciPy to calculate the determinate:

```
import numpy as np
from scipy import linalg
a = np.array([[1,4,2], [3,9,7], [8,5,6]])
print(linalg.det(a)) # calculate the matrix determinate
57.0
```

scikit-learn

scikit-learn is a machine learning library that is an extension of SciPy and is built using NumPy and Matplotlib. It contains many prebuilt machine learning models, such as random forests, K-means, and support vector machines. For more information on the different APIs it provides, view the documentation by visiting its website: https://scikit-learn.org/stable/user_guide.html.

In the following example, we will use an example dataset provided by scikit-learn and build a simple logistic regression model. First, we import all the required libraries and then load the Iris dataset provided by scikit-learn. We can use a handy API from scikit-learn to split our data into training and test datasets:

```
from sklearn import datasets
from sklearn.linear_model import LogisticRegression
from sklearn.model_selection import train_test_split
from sklearn.metrics import accuracy_score
import numpy as np
# Load the iris dataset
X, y = datasets.load_iris(return_X_y=True)
```

```
X_train, X_test, y_train, y_test = train_test_split(X, y, test_
size=0.20, random_state=1) Create linear regression object
```

We can then initialize our logistic regression model and simply run the `fit` function with our training data to train the model. Once we train our model, we can use it to make predictions and then measure its accuracy:

```
# Create Logistic Regression model
model = LogisticRegression()
# Train the model using the training sets
model.fit(X_train, y_train)
# Make predictions using the testing set
y_pred = model.predict(X_test)
print(accuracy_score(y_test, y_pred))
```

In the next few sections, we will start looking at the different models we can build using these libraries. We will understand what makes these models unique, how they are structured, and for what purposes and applications they can best serve our needs.

Building regression models

First, we will look at regression models. Regression models or regression analysis are modeling techniques used to find the relationship between independent and dependent variables. The output of a regression model is typically a continuous value, also known as a quantitative variable. Some common examples are predicting the price of a home based on its features or predicting the sales of a certain product in a new store based on previous sales information.

Before building a regression model, we must first understand the data and how it is structured. The majority of regression models involve supervised learning. This consists of features and an output variable, known as a label. This will help the model by adjusting the weights to better fit the data we have observed so far. We usually denote our features as X and our labels as Y to help us understand the mathematical models used to solve regression models:

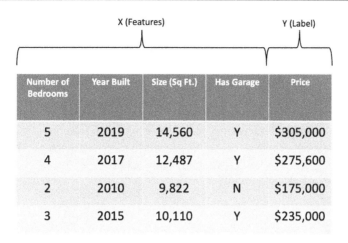

Number of Bedrooms	Year Built	Size (Sq Ft.)	Has Garage	Price
5	2019	14,560	Y	$305,000
4	2017	12,487	Y	$275,600
2	2010	9,822	N	$175,000
3	2015	10,110	Y	$235,000

Figure 4.8: Example of a supervised learning data structure

Typically, our data is split into two subsets, **training** and **testing** sets. The training dataset usually consists of between 70-80% of the original data and the testing dataset contains the rest. This is to allow the model to learn on the training dataset and validate its result on the testing dataset to show its performance. From the results, we can infer how our model is performing on the dataset.

For a linear regression model to perform effectively, our data must be structured linearly. The model uses this formula to train on and learn about the data:

$$y_i = \beta_0 + \beta_1 x_1 + \cdots + \beta_n x_n$$

In this equation, y_i represents the output of the model, or what we usually call the prediction. The prediction is calculated by taking the intercept β_0 and the slope β_n. The slope, which is also referred to as the weight, is applied to all the features in the data, which represents x_n. When working with the data, we usually represent it as a matrix, which makes it easy to understand and easy to work with when using Python:

$$y = X\beta$$

$$y = \begin{bmatrix} y_1 \\ y_2 \\ \cdot \\ \cdot \\ \cdot \\ y_n \end{bmatrix}$$

$$X = \begin{bmatrix} x_{11} & \cdots & x_{1c} \\ \vdots & \ddots & \vdots \\ x_{r1} & \cdots & x_{rc} \end{bmatrix}$$

$$\beta = \begin{bmatrix} \beta_0 \\ \beta_1 \\ \cdot \\ \cdot \\ \cdot \\ \beta_n \end{bmatrix}$$

The number of features describes what type of problem you are solving. If your data only has one feature, it is considered a simple linear regression model. While it can solve straightforward problems, for more advanced data and problems, it can be difficult to map relationships. Therefore, you can create a multiple linear regression model by adding more features (x_i). This allows the model to be more robust and find deeper relationships.

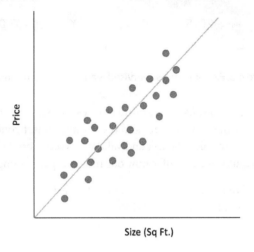

Figure 4.9: A simple linear regression model

Once we train our model, we need to learn how to evaluate our model and understand how it performs against the test data. When it comes to linear regression, the two common metrics we use to assess our model are the **Root Mean Square Error** (**RMSE**) and the **R**^{2} metrics.

The RMSE is the standard deviation of the **residual** errors across the predictions. The residual is the measure of the distance from the actual data points to the regression line. The further the average distance of all the points is from the line, the higher the error is. This indicates a weak model, as it's unable to find the correlation between the data points. This metric can be calculated by using this formula where y_i is the actual value, $\hat{y}_i$ is the predicted value, and N is the number of data points:

$$RMSE = \sqrt{\frac{\sum_{i=1}^{N}(y_i - \hat{y}_i)^2}{N}}$$

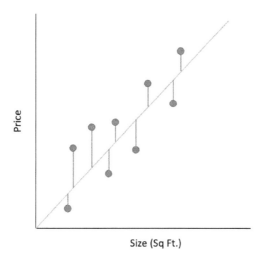

Figure 4.10: Calculating the residual of a linear regression model

R^2, also known as the coefficient of determination, measures the proportion of variance in the dependent variables (Y) that can be explained by the independent variables (X). It essentially tells us how well the data fits the model. Unlike the RMSE, which can be an arbitrary number, R^2 is given as a percentage, which can be easier to understand. The higher the percentage, the better the correlation of data. Although useful, a higher percentage is not always indicative of a strong model. What determines a good R^2 value depends on the application and how the user understands the data. R^2 can be calculated by using this formula:

$$R^2 = 1 - \frac{\sum_{i=1}^{N}(y_i - \widehat{y_i})^2}{\sum_{i=1}^{N}(y_i - \widehat{y_i})^2}$$

Many more metrics can evaluate the effectiveness of your regression model, but these two are more than enough to get an understanding of how your model is performing. When building and evaluating your model, it is important to plot and visualize your data and model, as this can identify key points. The plots can help you determine whether your model is **overfitting** or **underfitting**.

Overfitting occurs when your model is too suited to your training data. Your RMSE will be really low, and you will have a training accuracy of almost 100%. While this seems tempting, it is an indication of a poor model. This can be caused by one of two things: not enough data or too many parameters. As a result, when you test your model on new data it has not seen before, it will perform very poorly due to it not being able to generalize the data.

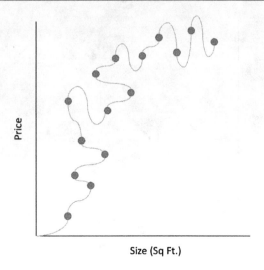

Figure 4.11: An overfitted linear regression model

To address overfitting, you can try to increase the amount of training data, or make the model less complex. It also helps to randomly shuffle your data before you split it into the training and testing set. Another important technique is called **regularization**. While there are many different regularization techniques (L1 or L2 regularization) depending on the model, they all work similarly in that they add **bias** or noise into the model to prevent overfitting. In the regression equation we previously saw, we can add another term, ϵ, to show that regularization is being applied to our model:

$$y_i = \beta_0 + \beta_1 x_1 + \cdots + \beta_n x_n + \epsilon$$

On the other end, underfitting occurs when your model is unable to find any meaningful correlation within the data. This is not as common as overfitting since it is easy to find patterns in most data. If this occurs, either your data has too much noise and is severely uncorrelated, your model is too simple and doesn't have enough parameters, or the model is not effective for the application at hand. It is also useful to debug your code and make sure there are no bugs when it comes to preprocessing your data or setting up your model:

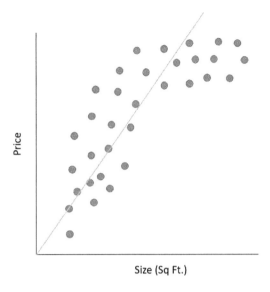

Figure 4.12: An under-fitted linear regression model

Therefore, the goal is to find the best-fitting model, between an overfit and an underfit. It takes time and experimentation to find a model that works for your needs, but using the key indicators and metrics discussed here can help guide you in the right direction:

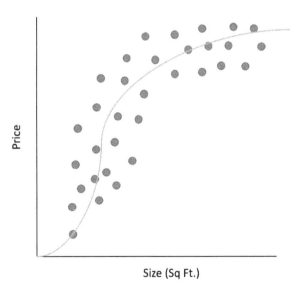

Figure 4.13: A best-fitting linear regression model

> **Important Note**
> Feature engineering is a critical part of building a comprehensive model. Understanding your data can help determine which features or parameters to include in your model so that you can capture the relationship between the independent and dependent variables without causing overfitting.

There are some key notes to keep in mind when collecting and working with the data for your model:

- **Normalize your data**: It is possible to have features with very high or low numbers, so to prevent them from overwhelming the model and creating biases, it is imperative to normalize all your data to make it uniform across the features.

- **Clean your data**: In the real world, the data we collect isn't always perfect and can contain missing or egregious data. It is important to deal with these issues because they can cause outliers and impact the model negatively.

- **Understand the data**: It is a common practice to perform statistical analysis, also known as **Exploratory Data Analysis** (**EDA**), on your data to get a better understanding of how the data can impact your model. This can include plotting graphs, running statical methods, and even using machine learning techniques to reduce the dimensionality of the data, which will be discussed later in the chapter.

In the next section, we will discuss classification models.

Building multiclass classification

Unlike regression models that produce a continuous output, models are considered classification models when they produce a finite output. Some examples include email spam detection, image classification, and speech recognition.

Classification models are considered versatile since they can apply to both supervised and unsupervised learning while regression models are mostly used for supervised learning. There are some regression models (such as logistic regression and support vector machine) that are also considered classification models since they use a threshold to split the output of continuous values into different categories.

Unsupervised learning is a common application used in today's market. Although supervised learning usually performs better and provides meaningful results since we know the expected output, the majority of the data we collect is unlabeled. It costs companies time and money for human experts to sift through the data and label it. Unsupervised learning helps reduce the cost and time by getting the model to try and determine the labels for the data and extract meaningful information. They can even perform better than humans sometimes.

The number of categories in the output of a classification model determines what type of model it is. For models with only two outputs (i.e., spam and not spam), this is called a binary classifier, while models with more than two outputs are called multiclass classifiers:

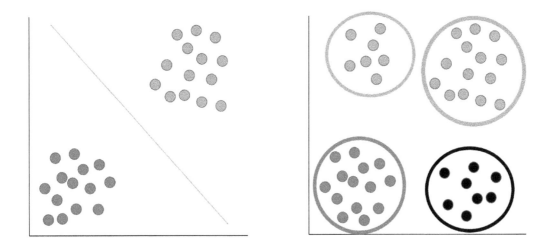

Figure 4.14: Binary and multiclass classifiers

From those classifiers, there are two types of learners: **lazy learners** and **eager learners**.

Lazy learners essentially store the training data and wait until they receive new test data. Once they get the test data, the model classifies the new data based on the already existing data. These types of learners take less time when training since you can continuously add new data without having to retrain the entire model, but take longer when performing classification since they have to go through all the data points. One common type of lazy learner is the **K-Nearest Neighbors** (**KNN**) algorithm.

On the other hand, eager learners work in the opposite way. Whenever new data is added to the model, they have to retrain the model again. Although this takes more time compared to lazy learners, querying the model is much faster since they don't have to go through all the data points. Some examples of eager learners are decision trees, naïve Bayes, and ANNs.

> **Important Note**
>
> Supervised learning will generally perform better than unsupervised learning since we know what the expected output should be during training, but it is costly to have to collect and label the data, so unsupervised learning excels in this area of training on unlabeled data.

In the next few sections, we will be looking at a few niche models that can be used for unique problems that most basic classification or regression models can't solve.

Text sentiment analysis and topic modeling

A popular field in the machine learning field is topic modeling and text analysis. With a plethora of text on the internet, being able to understand that data and create complex models such as chatbots and translation services has become a hot topic. Interacting with human language using software is called **NLP**.

Despite the amount of data we can use to train our models, it is a difficult task to create meaningful models. Language itself is complex and contains many grammar rules, especially when trying to translate between languages. Certain powerful techniques can help us when creating NLP models though.

> **Important Note**
> Before implementing any NLP models, it is imperative to preprocess the data in some way. Documents and text tend to contain extraneous data, such as stopping words (the/a/and) or random characters, which can affect the model and produce flawed results.

The first idea we will discuss is **topic modeling**. This is the process of grouping text or words from documents into different topics or fields. This is useful when you have a document or text and want to classify and group it into a certain genre without having to go through the tedious process of reading documents one by one. There are many different models used for topic modeling:

- **Latent Semantic Analysis (LSA)**
- **Probabilistic Latent Semantic Analysis (PLSA)**
- **Latent Dirichlet Allocation (LDA)**

We will focus on LDA. LDA uses statistics to find patterns and repeated occurrences of words or phrases and groups them into their topics. It assumes that each document contains a mixture of topics and that each topic contains a mixture of words. LDA first starts the process of going through the documents and keeping a word matrix, where it contains the count of each word in each document:

	House	Animal	Bills	Dog	City	Economy
Document 1	4	5	6	0	3	8
Document 2	0	8	5	5	2	0
Document 3	1	7	8	2	2	1
Document 4	2	6	7	1	1	0

Figure 4.15: A word matrix

After creating a word matrix, we determine the number of topics, k, that we want to split up the words into and use statistics to find the probability of the words belonging to a certain topic. Using Bayesian statistics, we can then calculate the probability and use that to cluster the words into different topics:

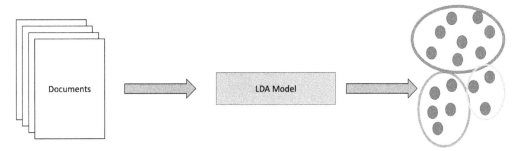

Figure 4.16: An LDA model

Another rising application in NLP is **sentiment analysis**. This involves the process of taking words or text and understanding the user's intent or emotion. This is common today when dealing with online reviews or social media posts. It determines whether a piece of text contains positive, neutral, or negative emotions.

Many different methods and models can solve this problem. The simplest approach is through statistics by using Bayes' theorem. This formula is used for predictive analysis, as it uses previous words in a text to update the model. The probability can be calculated using this formula:

$$P(A|B) = \frac{P(B|A)P(A)}{P(B)}$$

Deep learning has become a powerful tool for NLP and can be useful for sentiment analysis. CNNs and **Recurrent Neural Networks (RNNs)** are two types of deep learning models that can drastically improve models for NLP, especially for sentiment analysis. We will discuss these neural networks and how they perform more later in this chapter.

Pattern analysis and forecasting in machine learning

With the uncertainty of time, being able to predict certain trends and patterns has become a hot topic in today's industry. Most regression models, while powerful, are not able to make confident time predictions. As a result, some researchers have devised models that take time into consideration when making certain predictions, such as gas prices, stock market, and sales forecasting. Before we go into the different models, we must first understand the different concepts in time-series analysis.

The first step when dealing with time-series problems is familiarizing yourself with the data. The data usually contains one of four data components:

- **Trend** – The data follows an increasing or decreasing continuous timeline and there are no periodic changes

- **Seasonality** – The data changes in a set periodic timeline

- **Cyclical** – The data changes but there is no set periodic timeline

- **Irregular** – The data changes randomly with no pattern

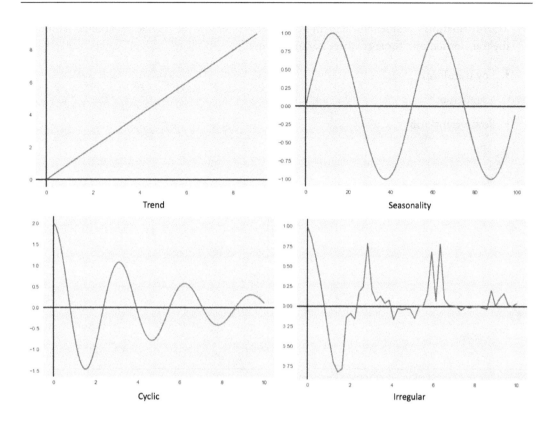

Figure 4.17: Different components of time-series data

These different trends can be split into two different data types:

- **Stationary** – Certain attributes of the data, such as mean, variance, and covariance, do not change over time

- **Non-stationary** – Attributes of the data change over time

Often, you will work with non-stationary data, and creating machine learning models using this type of data will generate unreliable results. To resolve this issue, we use certain techniques to change our data into stationary data. They include the following methods:

- **Differencing** – A mathematical method used to normalize the mean and remove the variance. It can be calculated by using this formula:

$$\hat{y}_t = y_t - y_{t-1}$$

- **Transformation** – Mathematical methods are used to remove the change in variance. Among the transformations, these are three commonly used:

  - Log transform

  - Square root

  - Power transform

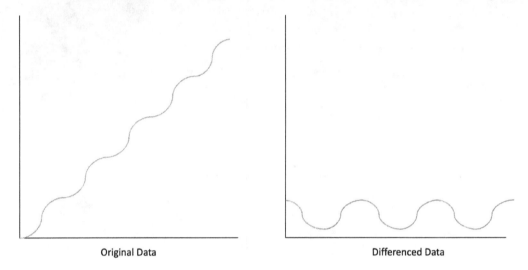

Original Data   Differenced Data

Figure 4.18: Differencing non-stationary data

Important Note

Time is already uncertain, and this makes it almost impossible to create a model that can confidently predict future trends. The more we can remove uncertainty in our data, the better our model can find the relationships in our data.

Once we can transform our data, we can start looking at models to help us with forecasting. Of the different models, the most popular model for time-series analysis is an **Auto-Regressive Integrated Moving Average (ARIMA)** model. This linear regression model consists of three subcomponents:

- **Auto-Regression (AR)** – A regression model that uses the dependencies of the current time and previous time to make predictions

- **Integrated (I)** – The process of differencing in order to make the data stationary

- **Moving Average (MA)** – Models between the expected data and the residual error by calculating the MA of the lagged observed data

Along with ARIMA, other machine learning models can be used for time-series problems. Another well-known model is the RNN model. This is a type of deep learning model used for data that has some sort of sequence. We will be going into more detail on how they work in the next section.

Enhancing models using deep learning

Earlier in the chapter, we briefly discussed deep learning and the advantages it brings when enhancing simple machine learning models. In this section, we will go into more information on the different deep learning models.

Before we can build our model, we will briefly go over the structure of the deep learning models. A simple ANN model usually contains about two to three fully connected layers and is usually strong enough to model most complex linear functions, but as we add more layers, the improvement to the model significantly diminishes, and it is unable to perform more complex applications due to overfitting.

Deep learning allows us to add multiple hidden layers to our ANN while reducing our time to train the model and increasing the performance. We can do this by adding one or both types of hidden layers common in deep learning – a CNN or RNN.

A CNN is mostly applied in the image detection and video recognition field due to how the neural network is structured. The CNN architecture comprises the following key features:

- Convolutional layers
- Activation layers
- Pooling layers

The convolutional layer is the core element in the CNN model. Its primary task is to convolve or group sections of the data using a **kernel** and produces an output called a **feature map**. This map contains all the key features extracted from the data, which can then be used for training in the fully connected layer. Each element in the feature map indicates a receptive field, which is used to denote which part of the input is used to map to the output. As you add more convolutional layers, you can extract more features, and this allows your model to adapt to more complex models:

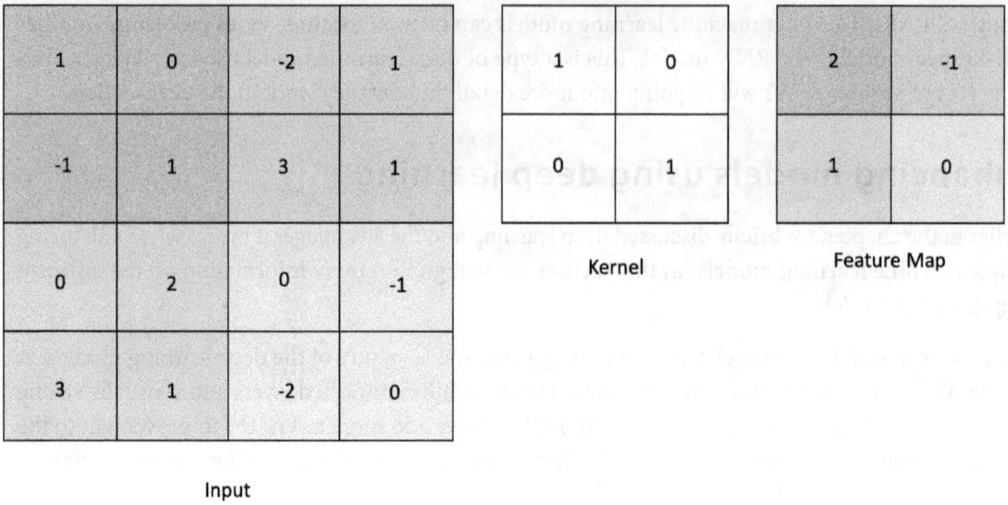

1	0	-2	1		1	0		2	-1
-1	1	3	1		0	1		1	0
0	2	0	-1						
3	1	-2	0						

Input

Kernel

Feature Map

Figure 4.19: Convolutional layer output

In *Figure 4.19*, we can see how the feature map is created in the convolutional layer. The layer slides the kernel across the input data, performs the dot operation, and produces an output matrix, which is the result of the convolution function. The size of the kernel and the step size can dictate the output size of the feature map. In this example, we use a kernel size of 2x2 and a stride or step size of 2, which gives us a feature map of size 2x2 based on our input size. The output of the feature map can then be used for more future convolutional layers depending on the requirements of the user.

Before we use our newly created feature map as an input for another convolutional or fully connected layer, we pass the feature map through an activation layer. It is important to pass your data through some type of nonlinear function during the training process, as this allows your model to map to more complex functions. Many different types of activation functions can be used throughout your model and have their benefits depending on the type of model you are planning to build. Among the many activation functions, these are the most commonly used:

- **Rectified Linear Activation (ReLU)**
- **Logistic (sigmoid)**
- The **Hyperbolic Tangent (Tanh)**

The ReLU function is the most popular activation function used today. It is very simple and helps the model learn and converge more quickly than most other activation functions. It is calculated using this function:

$$f(x) = \begin{cases} 0, & x < 0 \\ x, & x \geq 0 \end{cases}$$

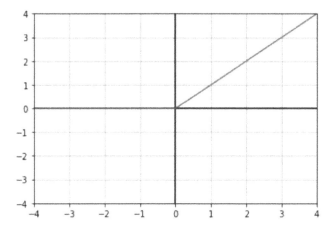

Figure 4.20: ReLU function

The logistic function is another commonly used activation function. This is the same function used in logistical regression models. This function helps bound the output of the feature map between 0 and 1. While useful, this function is computationally heavy and may slow down the training process. It is calculated using this function:

$$f(x) = \frac{1}{1 + e^{-x}}$$

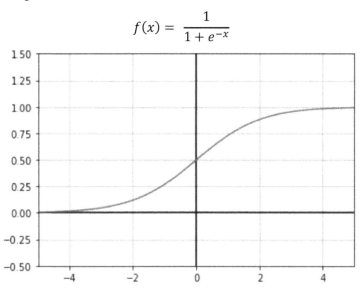

Figure 4.21: Sigmoid function

The Tanh function is similar to the sigmoid function in that it bounds the values from the feature map. Rather than bounding it from 0 to 1, it bounds the values from -1 to 1, and it usually performs better than a sigmoid function. The function is calculated using this formula:

$$f(x) = \frac{e^x - e^{-x}}{e^x + e^{-x}}$$

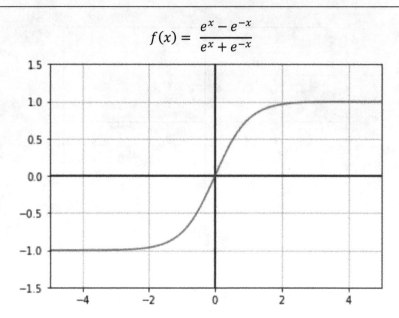

Figure 4.22: Tanh function

Each activation has its uses and benefits depending on the task or model at hand. The ReLU function is commonly used in CNN models while sigmoid and Tanh are mostly found in RNN models, but they can be used interchangeably and bring different results.

After we run our feature map through an activation layer, we come to the final piece – the pooling layer. As mentioned before, a key element in deep learning is the reduction of parameters. This allows the model to train on fewer parameters while still retaining the important features extracted from our convolutional layers. The pooling layer is responsible for this step of downsizing our parameter size. There are many common pooling functions but the most commonly used is max pooling. This is similar to the convolutional layer, where we use a kernel or filter to slide through our input data and only take the maximum value from each window:

1	0	-2	1
-1	1	3	1
0	2	0	-1
3	1	-2	0

Input

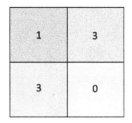

Feature Map

Figure 4.23: Max pooling layer output

In *Figure 4.23*, we are using a kernel size of 2x2 with a stride size of 2. Here, we can see our output where only the maximum value from each window is selected.

Other layers and functions can be added to the model to help address certain issues and applications, such as a batch normalization layer, but with these three foundational layers, we can add multiple layers of different sizes and still build a powerful model.

In the final layer, we feed our output into the fully connected layer. By that time, we are able to extract the important features of the data and still learn more complex models with less time to train as compared to a simple ANN model.

Next, we will go over the RNN architecture. Due to the nature of how the RNN model is structured, it is designed for tasks that need to take into consideration a set of sequence data, in which the data later in the sequence is dependent on earlier data. This model is commonly used for certain fields such as NLP and signal processing.

The basics of an RNN are built by having a hidden layer in which the output of the layer is fed back into the same hidden layer. This way, the model is able to learn based on previous data and adjust the weights accordingly.

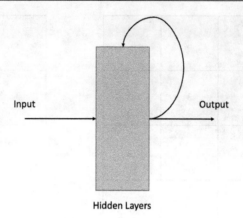

Figure 4.24: RNN architecture

To better understand how the model works, you can envision a single layer for each data point. Each layer takes in the data point as an input x_i and produces an output y_i. We then transfer the weights between the layers and then take the total average of all the cost functions from all the layers in the model.

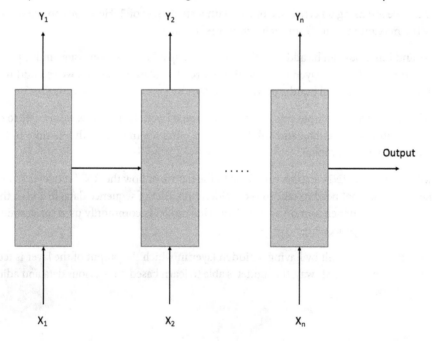

Figure 4.25: An unraveled RNN

This model works for most simple problems. As the sequence increases, it encounters a few issues:

- **Vanishing gradient**: This occurs during the training process when the gradient approaches zero. The weights aren't updated properly as a result and the model performs poorly.

- **Lack of context**: The model is unidirectional and cannot look further or previously into the data. Therefore, the model is only able to predict based on data around the current sequence point and is more likely to make a poor prediction based on incorrect context.

There are different variations of RNNs created to address some of the issues mentioned here. Among them, the most common one used today is the **Long Short-Term Memory** (**LSTM**) model. The LSTM model comprise three components:

- An input gate
- An output gate
- A forget gate

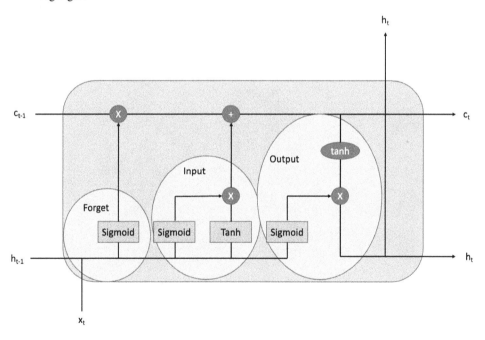

Figure 4.26: An LSTM neural network

These gates work by regulating which data points are needed to contextualize the sequence. That way, the model can predict more accurately without being easily manipulated.

The forget gate is specifically responsible for removing previous data or context that is no longer needed. This gate uses the sigmoid function to determine whether it uses or "forgets" the data.

The input gate is used to determine whether the new data is relevant to the current sequence or not. This is so that only important data is being used to train the model and not redundant or irrelevant information.

Lastly, the output gate's primary function is to filter the current state's information and only send relevant information to the next state. As with the other gates, it uses the context from previous states to apply a filter, which helps the model properly contextualize the data.

CNN and RNN models are mostly designed for supervised learning problems. When it comes to unsupervised learning, different models are needed to solve certain problems. Let's discuss **autoencoders**.

Autoencoders work by taking the input data, compressing it, and then reconstructing it by decompressing it. While straightforward, it can be used for some advanced applications, such as generating audio or images, or it can be used as an anomaly detector.

The autoencoder comprises two parts:

- An encoder
- A decoder

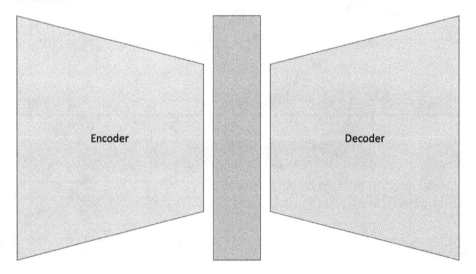

Figure 4.27: The components of an autoencoder

The encoder and decoder are usually built with a one-layer ANN. The encoder is responsible for taking the data and compressing or flattening the data. Then, the decoder works on taking the flattened data and trying to reconstruct the input data.

The hidden layer in the middle of the encode and decoder is usually referred to as the bottleneck. The number of nodes in the hidden layer must be less than those in the encoder and decoder. This forces the model to try and find the pattern or representation in the input data so that it can reconstruct the data with little information. Thus, the cost function is there to calculate and minimize the difference between the input and output data.

One aspect of autoencoders that is an integral part of deep learning is **dimensionality reduction**. This is the process of reducing the number of parameters or features used when training your model. As mentioned earlier in this chapter, to build a complex model that can build a deeper representation of the data, it is important to include more features. However, adding too many features can lead to overfitting, so how do we find the best number of features to use in our model?

There are many models and techniques, such as autoencoders, that can perform dimensionality reduction to help us find the best features to use in our model. Among the different techniques, **Principle Component Analysis (PCA)** is the most popular. This technique can take an N-dimensional dataset and reduce the number of dimensions in the data using linear algebra. It is a common practice to use a dimensionality reduction technique before using your data to train your model, as this can help to remove noise in the data and avoid overfitting.

Summary

In this chapter, we discussed what is considered artificial intelligence and the different sub-fields that it contains.

We discussed the different characteristics of regression and classification models. From there, we also went over the structure of our data and how the model performs when training over our data. We then discussed the different ways of analyzing our model's performance and how to address the different issues that we can come across when training our model.

We briefly viewed the different packages and libraries that are used today in machine learning models and their different use cases.

We also analyzed different topics such as topic modeling and time-series analysis and what they entail. With that, we were able to look at the different methods and techniques used to solve those types of problems.

Lastly, we went into deep learning and the different ways it improves on machine learning. We went over the two different types of neural networks – CNNs and RNNs – how they are structured, and their benefits and use cases.

In the next chapter, we will take what we have learned and start looking into how we can design and build an end-to-end machine learning system and the different components that it contains.

5

Machine Learning System Design

In the last chapter, we delved into the different machine learning concepts and the packages and libraries used to create these models. Using that information, we will begin to discuss the design process when building a machine learning pipeline and the different components found in most machine learning pipelines.

We will cover the following areas in this chapter:

- Machine learning system components
- Fit and transform interfaces
- Train and serve interfaces
- Orchestration

Machine learning system components

There are many moving parts required in order to build a robust machine learning system. Starting from gathering data to deploying your model to the user, each plays a vital role in keeping the system dynamic and scalable. Here, we will briefly discuss the different stages in the machine learning system life cycle and the role they play. These stages can be edited in order to suit the model or application at hand.

The majority of machine learning systems include the following stages, with some other stages depending on business needs:

- **Data collection**
- **Date preprocessing**
- **Model training**

- **Model testing**

- **Model serving**

Realistically, the majority of the time spent building machine learning systems is spent on the data. This is a key element in the process that can decide the effectiveness of your system since the model is dependent on the data it uses during training. Just like the human body, if you feed the model poor data or not enough data, it will output poor results.

The first part when it comes to data is the collection process. Understanding the application and the goal of the task can assist in the process of deciding how to collect data and what data to collect. We then determine the target value that we want to predict, such as the price of a home or the presence of a certain disease. These target values can be collected explicitly or implicitly. A target variable is explicit when we can directly determine the value of the variable we are trying to capture, while an implicit target value is found by using contextual data to determine the target value.

Depending on the task, we usually store the data in a database (for either metadata or tabular data) such as MySQL or cloud storage (for images, video, or audio) such as Amazon S3:

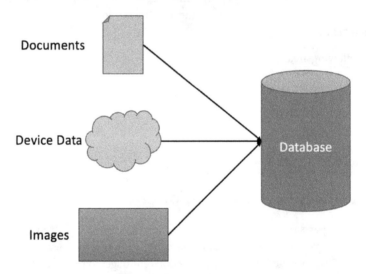

Figure 5.1: Data collection

Once we set up continuous data collection, we must devise a procedure for cleaning and processing the data. Not everything we collect will be perfect. You will always find missing data and certain outliers, which can negatively impact our model. No matter how intuitive your model is, it will always perform poorly with garbage data.

Some practices to deal with unclean data include removing outliers, normalizing certain features, or imputing missing data depending on the amount of data you have collected. Once the data has gone through the cleaning process, the next step is the feature selection/engineering process.

Understanding the different features your data contains plays an important role when your model tries to find the relationship in its data. **Exploratory Data Analysis** (**EDA**) is the common process used when it comes to understanding the data you have collected and how the data is structured. This helps when it comes to determining which features to use in your model. As we previously mentioned in *Chapter 4*, when we include more features in our models, it allows them to map to more complex problems. However, adding too many features can lead to overfitting, so it is important to research the most important features for your model.

While most machine learning models can find patterns and relationships in data, the best way of understanding the data you collect is via the experts in the field of the task you are trying to solve. Subject matter experts can provide the best insight into what features to focus on when creating your model. Some unsupervised machine learning models, such as PCA and t-SNE, can group and find features that can provide the most valuable information for your model.

> **Important note**
> Having domain knowledge of the problem you are trying to solve is the most effective way of understanding your data and determining which features to use for training your machine learning model.

Once you have set up the processes to collect and clean the data, the next step is creating and training your model. Thanks to most machine learning libraries, you can import prebuilt models and even use weights from already trained models to use on your own model. Here, it is common practice to use different models and techniques to see which produces the best result, and from there, you can choose the best model and begin to fine-tune it by updating the hyperparameters. This process can take time depending on the amount of data you use.

Testing your model is a critical element in your system's pipeline. Depending on the application, a poor model can negatively impact your business and give your users a bad experience. To prevent that, you need to determine the different metrics and thresholds that need to be met for the model to be production-ready. If the model can't meet these expectations, then you need to go back and understand the weaknesses of the model and address them before training again.

After performing tests and getting solid results from your model, you can now deploy your model to the user application. This varies from application to application. From then, the whole process can start from the beginning, where new data is inserted and follows the machine learning pipeline so it can dynamically grow based on user actions:

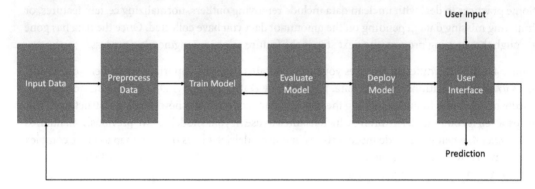

Figure 5.2: The machine learning pipeline

In the following sections, we will look into the details of the different interfaces that constitute our machine learning pipeline.

Fit and transform interfaces

Now that we have looked at the entire pipeline process, we will look in detail at the different interfaces that make up the machine learning system. The majority of the systems include the following interfaces:

- **Fit**
- **Transform**
- **Train**
- **Serve**

When it comes to the data and creating the model, we come across the fit and transform interfaces. We will start by looking at the transform interface.

Transform

The transform interface is the process of taking in the collected data and preprocessing the data so that the model can train properly and extract meaningful information.

It is common for the data we collect to have missing values or outliers, which can cause bias in our model. To remove this bias, we can apply certain techniques that help remove the skew in the data and produce meaningful machine learning models. Some of the following techniques we will learn about fall into the following three types of transformations:

- **Scaling**

- **Clipping**

- **Log**

Log transformation is the most common and simple transformation technique we can apply to our data. A lot of the time, our data is skewed in one direction, which can introduce bias. To help mitigate the skewed distribution, we can simply apply the log function to our data, and this shifts our data into more of a normal distribution, which allows the data to be more balanced.

We can perform this transformation by using the following code:

```
import numpy as np
dataframe_log = np.log(dataframe["House Price"])
```

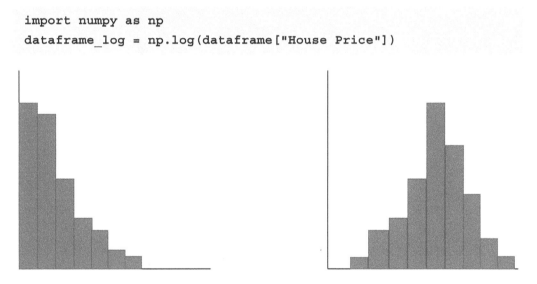

Figure 5.3: Performing log transformation on skewed data

Once we apply the log transformation, we can start looking at the other transformations. The second transformation we can use is the clipping transformation. The more we make our data follow a normal distribution, the better, but we may encounter outliers that can skew our data. To help reduce the impact that outliers have on our data, we can apply a quantile function. The most common quantile range that people use is the 0.05 and 0.95 percentile. This means that any data below the 0.05 percentile will be rounded up to the lower bound while any data above the 0.95 percentile will be rounded down to the upper bound. This allows us to retain the majority of the data while reducing the impact that outliers have on the model. The upper and lower ranges can also be modified based on what makes sense for the distribution of the data.

This transformation can be performed using the following code:

```
from sklearn.preprocessing import QuantileTransformer
quantile = QuantileTransformer(output_distribution='normal',
random_state=0)
x_clipped = quantile.fit_transform("House Price")
```

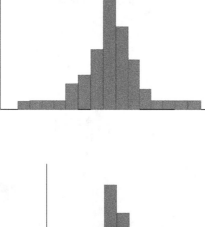

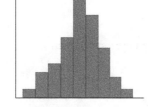

Figure 5.4: Clipping transformation on data

The last major transformation technique is scaling transformations. A lot of the time, the data we collect have different types of metrics and values, which can skew our data and confuse our model. For example, one feature measures the revenue of companies in the millions while another feature measures the employee count in the thousands, and when using these features to train the model, the discrepancy may put more emphasis on one feature over another. To prevent these kinds of problems, we can apply scaling transformations, which can be of the following types:

- **MinMax**
- **Standard**
- **Max Abs**
- **Robust**
- **Unit Vector**

The MinMax scaler is the simplest scaling transformation. It works best when the data is not distorted. This scales the data between 0 and 1. It can be calculated using this formula:

$$x_{scaled} = (x - x_{min}) / (x_{max} - x_{min})$$

We can perform this scaling transformation using the following code:

```
from sklearn.preprocessing import MinMaxScaler
scaler = MinMaxScaler()
x_scaled = scaler.fit_transform("House Price")
```

The MaxAbs scaler is similar to MinMax but rather than scaling the data between 0 to 1, it scales the data from -1 to 1. This can be calculated using the following formula:

$$x_{scaled} = \frac{\max(|x|)}{x}$$

We can perform this scaling transformation using the following code:

```
from sklearn.preprocessing import MaxAbsScaler
scaler = MaxAbsScaler()
x_scaled = scaler.fit_transform("House Price")
```

The Standard scaler is another popular scaling transformation. Rather than using the min and max like the MinMax scaler, this scales the data so that the mean is 0 and the standard deviation is 1. This scaler works on the assumption that the data is normally distributed. This can be calculated using the following formula:

$$x_{scaled} = x - \frac{\mu}{\sigma}$$

We can perform this scaling transformation using the following code:

```
from sklearn.preprocessing import StandardScaler
scaler = StandardScaler()
x_scaled = scaler.fit_transform("House Price")
```

The MinMax, MaxAbs, and Standard scalers, while powerful, can suffer from outliers and skewed distribution. To remedy this issue, we can use the Robust scaler. Rather than using the mean or max, this scaler works by removing the median from the data and then scaling the data using the interquartile range. This can be calculated using the following formula:

$$InterQuartile\ Range\ (IQR) = Q_3 - Q_1$$

$$x_{scaled} = (x - Q_1)/IQR$$

We can perform this scaling transformation using the following code:

```
from sklearn.preprocessing import RobustScalar
scaler = RobustScalar()
x_scaled = scaler.fit_transform("House Price")
```

Finally, we have the Unit Vector scaler, also known as a normalizer. While the other scaler functions work based on columns, this scaler normalizes based on rows. It uses the MinMax scaler formula and converts positive values between 0 and 1 and negative values between -1 and 1. There are two ways of performing this scaling:

- L1 norm – values in the column are converted so that the sum of their absolute value in the row equals 1

- L2 norm – values in the column are squared and added so that the sum of their absolute value in the row is equal to 1

We can perform this scaling transformation using the following code:

```
from sklearn.preprocessing import Normalizer
scaler = Normalizer()
x_scaled = scaler.fit_transform("House Price")
```

There are many more scaling and transforming techniques, but these are the most commonly used, as they provide stable and consistent results.

> **Important note**
> Much of the development process takes place in the transformation stage. Understanding how the data is structured and distributed helps dictate which transformation methods you will perform on your data. No matter how advanced your model is, poorly structured data will produce weak models.

Fit

Now, we will look at the fit interface. This interface refers to the process of creating the machine learning model that will be used in training. With today's technology, not much work or effort is needed to create the model used for training in the machine learning pipeline. There are already prebuilt models ready to be imported and used for any type of application.

Here is a small example of creating a KNN classification model using the scikit-learn library.

First, we import all the required libraries:

```
from sklearn.model_selection import train_test_split
from sklearn.preprocessing import StandardScaler
from sklearn.neighbors import KneighborsClassifier
from sklearn.datasets import load_iris
```

We then import the data, split the data into training and testing batches, and apply a standard scaler transformation:

```
iris = load_iris()
X = iris.data
y = iris.target
X_train, X_test, y_train, y_test = train_test_split(X, y, test_
size=0.30)
scaler = StandardScaler()
scaler.fit(X_train)
X_train = scaler.transform(X_train)
X_test = scaler.transform(X_test)
```

We then initialize a KNN model with k = 3 and then perform training on the model:

```
classifier = KNeighborsClassifier(n_neighbors=3)
classifier.fit(X_train, y_train)
```

The main effort when using the fit interface is setting up the models that will be used for the training phase of the machine learning pipeline. Due to the simplicity of importing multiple prebuilt models, it is common practice to import multiple types of machine learning models and train all of them at once. This way, we are able to test different types of models and determine which one of them performs the best. Once we decide which model to use, we can then start to experiment with different hyperparameters to further fine-tune our model.

Train and serve interfaces

The transform and fit interfaces are responsible for preparing the data and setting up our machine learning models for our pipeline. Now that we have preprocessed the data, we need to begin looking at how we can begin the actual training process and take our trained models and deploy them for our clients to use.

Training

Now that we have preprocessed the data and created our models, we can begin the training process. This stage can vary from time to time depending on the quality of data being trained on or the type of model being used during training.

Once we preprocess the data, we need to split the dataset into training and testing sets. This is done to prevent overfitting. We need the model to be able to generalize the data, and using all the data for training would defeat the purpose.

A common practice is to split your data into 70% training and 30% testing. This way, the model has enough data to learn the relationships and uses the testing data to self-correct its training process.

There is a more robust approach to splitting the data, which is called **K-Fold Cross-Validation**. This process works best in cases where there may not be enough training data. To perform this, we split the data into k number of subsets and then we train on all subsets except for one. We then iterate through this process where a new subset is selected to be the test data. Finally, we measure the performance of the model by averaging the metrics for each iteration. This way, we can train and test using all the data without leaving any important features that may be useful when it comes to learning the data.

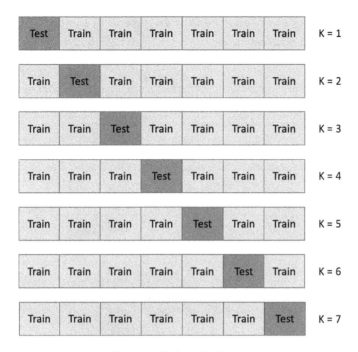

Figure 5.5: K-Cross Validation

Once we have split the data, now comes the actual training part. This part is as simple as setting up the function used to train the model. This part depends on the type of library you use and the different APIs it offers.

We can create a simple example using the scikit-learn library:

```
from sklearn.datasets import load_diabetes
from sklearn.model_selection import train_test_split
from sklearn.linear_model import LinearRegression
from sklearn import metrics

diabetes = load_diabetes()
features = diabetes.data
target = diabetes.target

x_train, x_test, y_train, y_test = train_test_split(features,
target, test_size=0.3, random_state=1)
linear_regression = LinearRegression()
linear_regression.fit(features, target)
```

```
y_pred = linear_regression.predict(x_test)
print("Linear Regression model MSE:", metrics.mean_squared_
error(y_test, y_pred))
```

After training your model, you must measure its performance. To prevent poor models from being deployed to users, it is a common practice to measure certain metrics and set certain thresholds that need to be met before a model is considered ready for production.

Depending on the type of model you create, certain metrics need to be evaluated. For example, a regression model will typically look at the following metrics:

- **Mean Absolute Error (MAE)**
- **Mean Squared Error (MSE)**
- **Root Mean Squared Error (RMSE)**
- **R-Squared (R2)**

For classification models, you will monitor the following metrics to determine the model's strength:

- Accuracy
- Precision and recall
- The F1-score
- The **Area Under the Receiver Operating Characteristics Curve (AUROC)**

Having domain knowledge helps immensely when determining what thresholds are applicable to the model you are training. In some cases, such as with cancer detection models, it is important to avoid false negatives, so it is important to set stricter thresholds for what models can be used confidently.

> **Important note**
>
> Before serving your model, you need to make sure the model is viable for production. Setting up the metric thresholds that the model needs to pass is a fundamental way of validating your models before deploying them. If your model fails to pass these criteria, then there should be a process to redo the data transformation and model training phases until it can pass the thresholds.

Serving

When it comes to serving our model, this is open and flexible depending on the user's needs. In most cases, we are deploying our model into one of two types of systems:

- **Model serving**, where we deploy our model as an API
- **Model embedding**, where we deploy our model straight into an application or device

Model embedding is the simplest way of deploying your model. You create a binary file containing your model and you embed the file into your application code. This simplicity provides the best performance when making predictions, but this comes at a cost. Because you directly embed the file into your application, it is difficult to scale your model since you will have to recreate and reupload the file every time you make an update to your model. As such, this is not considered a recommended practice.

Model serving is the most commonly used method on today's market. This separation between the application and the model makes it easy for a developer to maintain and update the model without having to change the application itself. You simply create an API service that a user can access to make calls and predictions. Due to the separation, you can continuously update the model without having to redeploy the whole application.

An alternative to model embedding that includes model serving is creating a microservice that includes the binary file of the model, which could be accessed by other applications:

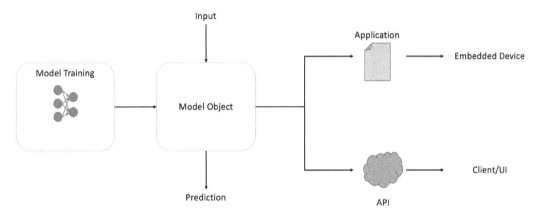

Figure 5.6: Serving machine learning models

One of the more intuitive approaches is creating your own package or library that includes all the models that you have trained. That way, you can scale efficiently by allowing multiple applications to access the different models you have created.

Everything we've seen so far is what it takes to build a simple machine learning pipeline. While this is doable for most applications, to be dynamic and robust, we need to look at orchestration and what it can offer us to support more advanced applications and problems.

Orchestration

Now that we understand the different interfaces and the roles they play in the machine learning pipeline, the next step is understanding how to wrap everything together into one seamless system. To understand the holistic system, we must first understand automation and orchestration.

Automation refers to the process of automating small or simple tasks, such as uploading files to a server or deploying an application, without human intervention. Rather than having a person perform these repetitive tasks, we can program our system to handle these simple tasks, thus reducing wasted time and resources.

This is useful for most systems due to the linear nature of the pipeline. This highlights a common limitation of automation though – the lack of flexibility. Most systems today require a more dynamic process to be able to adapt to certain applications and processes, and automation alone isn't enough:

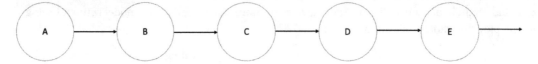

Figure 5.7: A linear system pipeline

This is where orchestration comes into action. **Orchestration** is the configuration and coordination of automated tasks to create a whole workflow. We can create a system to perform certain jobs or tasks based on a certain set of rules. It takes some planning and understanding to create a comprehensive orchestration workflow since the user determines what actions the system needs to take for certain cases.

A simple example would be deploying an application to users. There can be many moving parts in the system, such as the following:

- Connecting to a server
- Uploading certain files to certain servers
- Handling user requests
- Storing data or logs in a database

Let's say that after the recent changes have been deployed, the app has suffered critical errors, which may bring down the application. The system admin could set up rules for recovering and restoring the system, such as rolling back to a stable version. With the system able to self-recover, the developers can spend more time in development rather than dealing with overhead when it comes to recovery.

Depending on certain outcomes, not all tasks may need to be performed. There may be backup actions that need to take place, or different paths that the system needs to go through to maintain a stable workflow. This way, the system can adapt to its environment and self-sustain without much human intervention:

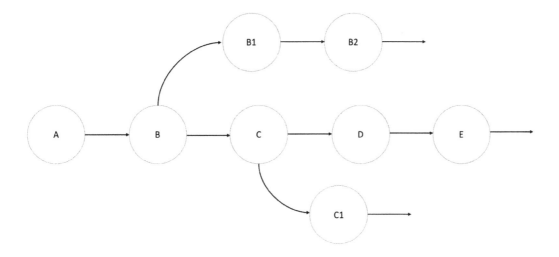

Figure 5.8: A dynamic system pipeline (orchestration)

The different tasks in the machine learning system that can be automated are as follows:

- Gathering and preprocessing the data
- Training the machine learning model
- Running tests and diagnostics on the trained model to evaluate its performance
- Serving the machine learning model
- Monitoring the model in production

With these automated tasks, the system admin needs to orchestrate the stages of the pipeline to be dynamic and sustainable. The following components help create a robust system:

- **Scheduling**: The system must be able to schedule and run different automated tasks in the pipeline individually while maintaining system dependencies.
- **CI/CD Testing**: After model training is complete, it is imperative to do automated testing on your model to measure its performance. If it fails to pass certain metrics, you must repeat the training process from the beginning to address the weaknesses of the model; otherwise, it cannot be deployed to production.
- **Deployment**: Depending on where you will deploy your model to production, setting up an automated process can help reduce the time spent on deployment and still maintain an updated version of the model.

- **Monitoring**: After deploying your model, continuously monitoring the model's performance in production is needed to maintain the model's health without it decaying. This will give us an indication of when we need to update our pipeline or our model in order to stay efficient.

> **Important note**
>
> Understanding what your business needs are and how your model functions gives you a good picture of how you want to orchestrate your machine learning pipeline. Setting up backup phases to address certain pitfalls in your system allows it to be more dynamic and adaptable to industry demands.

Summary

In this chapter, we looked at the different key components that make up a machine learning pipeline.

From there, we looked in detail at the interfaces that make up the components. We started with the transform interface, which is responsible for the data aspect of the pipeline. It takes the data and applies different types of data transformation that allow us to maintain clean and stable data, which we can later use in our machine learning model.

After our transformation stage, we start creating our model in the fit interface. Here, we can use the prebuilt models that the libraries and packages offer to initialize our models. Due to the ease of creating models, it is a good practice to test different types of models to see which model performs the best based on our data.

Once we have created our model, we can begin the actual training of our model. We need to split our data into training and test sets to allow our model to understand the relationship in our data. From there, we can measure the different metrics in our model to validate the model's performance.

Once we feel comfortable with our model's performance, we can start to deploy our application to production. There are two major ways of deploying our model, whether it be embedded into our application or deployed as a service for our clients to use.

Finally, wrapping everything together, we learned what orchestration consists of when it comes to machine learning. We learned what concepts need to be considered when orchestrating your machine learning pipeline and how to keep your system dynamic and robust to keep up with everyday demands.

As time passes and data changes, it is important that we adjust and maintain our models to handle certain situations that may arise in the real world. In the next chapter, we will look at how we can maintain our machine learning models when our data starts to shift and change.

6
Stabilizing the Machine Learning System

In the last two chapters, we went over the different concepts in machine learning and how we can create a comprehensive machine learning system pipeline that can work and adapt to our needs.

While our pipeline can address our expectations, it is important for us to be able to maintain our system in the face of external factors to which it may be hard for the system to self-adjust.

In this chapter, we will discuss the phenomenon of dataset shifts and how we can optimize our machine learning system to help address these issues while maintaining its functional goal without having to rebuild our system from scratch.

We will be going over the following concepts:

- Machine learning parameterization and dataset shifts
- The causes of dataset shifts
- Identifying dataset shifts
- Handling and stabilizing dataset shifts

Machine learning parameterization and dataset shifts

Maintaining our machine learning models is an integral part of creating a robust model. As time progresses, our data begins to morph and shift based on our environment, and while most models can detect and self-repair, sometimes, human intervention will be required to guide them back on track.

In this section, we will briefly go over two main concepts that will help us understand the impact on our model:

- **Parameterization**
- **Dataset shifts**

Our machine learning model is represented by certain specifications that help define the learning process of our model. These include the following:

- **Parameters**
- **Hyperparameters**

We will first look at parameters. These specifications are internal within the model. During the training process, these parameters are updated and learned while the model is trying to learn the mapping between the input features and the target values.

Most of the time, these parameters are set to an initial value of either zeros or random values. As the training process happens, the values are continuously updated by an optimization method, such as gradient descent. At the end of the training process, the final weights of the values are what constitute the model itself. These weights can even be used for other models, especially those with similar applications.

Some examples of parameters include the following:

- Node weights and bias values for artificial neural networks
- Coefficients of linear and logistic regression models
- Cluster centroids for clustering models

While parameters play a core role in determining the performance of a model, they are mostly out of our control since the model itself is what updates the weights. This leads us to hyperparameters.

Hyperparameters are parameters that control the learning process of our machine learning model, which, in turn, affects the output weights that our model learns. These values are set from the beginning and stay fixed throughout the learning process.

We, as users, determine which values to set in the beginning for our model to use during the training process. As a result, it takes time and experience to figure out which values produce the best results. There is effort involved in testing and training multiple variations of hyperparameters to see which performs the best.

There are many hyperparameters and each model has its own unique set of hyperparameters that the user can modify. These hyperparameters can include the following:

- The split ratio between the training and testing datasets
- The learning rate used in optimization algorithms
- The choice of optimization algorithm
- The batch size
- The number of epochs or iterations
- The number of hidden layers

- The number of nodes in each hidden layer

- The choice of cost or loss function

- The choice of activation function

- The number of K clusters

Since there can be many hyperparameters to adjust and many different combinations to try, it can be very time-consuming to test these changes one by one. As discussed in the last chapter, it can be useful to have a section in our pipeline that automates this process by running multiple models with different combinations of hyperparameters to speed up the testing process and find the most optimal combination of hyperparameters.

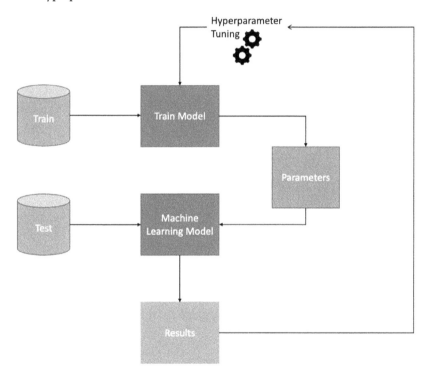

Figure 6.1: Hyperparameter and parameter tuning

There may be cases where adjusting our parameters and hyperparameters is not enough for us to prevent our model from degrading.

For example, let's say we create a machine learning model with a model accuracy of 85%. This model continues to perform well for some time. We then begin to see our model accuracy deteriorate until it becomes unusable, as the model is unable to properly predict the new test data we collect.

As we analyze our model, we can begin to see that our training data does not reflect the testing data we have recently collected. Here, we can see that there is a shift between the data distribution for our training and test datasets.

Before we work on resolving dataset shifts, we must first understand the background of dataset shifts, how they occur, and how we can adjust our machine learning system to help prevent dataset shifts from impacting our model.

Machine learning systems are built under the assumption that the data distribution between the training and test sets is similar. Since the real world is ever-changing, new data distributions emerge and there may be a significant difference between the training and test sets.

The major difference in data distribution between the training and test sets is considered a dataset shift. This drastic difference will eventually degrade the model, as the model is biased to the training set and is unable to adapt to the test set:

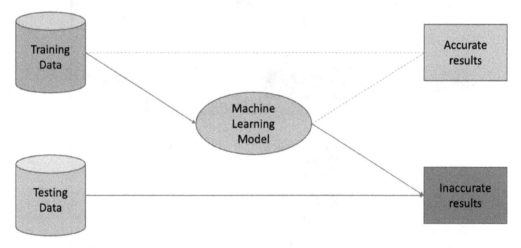

Figure 6.2: Outcome of a machine learning model due to a dataset shift

Some examples of this occurring include a shift in consumer habits, a socioeconomic shift, or a global influence, such as a pandemic. These events can heavily impact the data we collect and observe, which, in turn, can sway our model's performance.

> **Important Note**
>
> First, try adjusting the hyperparameters of your machine learning model and see whether the newly learned parameters can improve your model significantly. If you still encounter major issues, it may be best to analyze the data and see whether a dataset shift has occurred.

The causes of dataset shifts

Now that we have learned what dataset shifts are, we can start to investigate the different causes of dataset shifts. While there are many different reasons dataset shifts can occur, we can split them into two categories:

- **Sample selection bias**
- **Non-stationary environments**

Sample selection bias is self-explanatory in that there is a bias or issue when it comes to labeling or collecting the training data used for the model. Collecting biased data will result in a non-uniform sample selection for the training set. That bias, in essence, will fail to represent the actual sample distribution.

Non-stationary environments are another cause for dataset shifts – we will go into further detail about the different types later in the chapter. Let's assume that we have a model with a set of input features, x, a target or output variable y. From there, we can also define the prior probability as $p(x)$, the conditional probability as $p(y|x)$, and the joint distribution as $P(y, x)$. This dataset shift is caused by temporal or spatial changes, defined as $P_{train}(y, x) \neq P_{test}(y, x)$, which reflect very much how the real world operates.

This causal effect can lead to different types of shifts:

- For $X \rightarrow Y$ problems, non-stationary environments can make changes to either $p(x)$ or $P(y|x)$, giving us a covariate or concept shift
- For $Y \rightarrow X$ problems, a change in $p(y)$ or $P(x|y)$ can give us a prior probability or concept shift

In the next section, we will look into the different types of shifts and how we can identify them.

Identifying dataset shifts

After looking into the different causes of dataset shifts, we can begin to classify certain shifts into different groups that can help us easily identify the type of dataset shift we are dealing with.

Among the different dataset shifts we can encounter, we can classify data shifts into these categories:

- **Covariate shifts**
- **Prior probability shifts**
- **Concept shifts**

We will first look at covariate shifts. This is the most common dataset shift, as a covariate shift occurs when there is a change in the distribution of one or more of the input features of the training or test data. Despite the change, the target value remains the same.

In mathematical terms, this dataset shift occurs only in X > Y problems. Whenever the input distribution, $p(x)$, changes between the training and testing datasets, $p_{train}(x) \neq p_{test}(x)$, but the conditional probability of the training and testing dataset stays the same, $p_{train}(y|x) = p_{test}(y|x)$, this will cause a covariate shift.

For example, we can create a model that predicts the salary of the employees of a certain city. Let's say that the majority of the employees in your training set consist of younger individuals. After time passes, the employees get older. If you were to try to predict the salary of the older employees, you would begin to see a significant error. This is due to the model being heavily biased toward the training set, which consisted of mostly younger employees and is unable to find the relationship among the older employees.

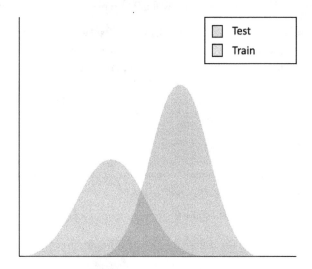

Figure 6.3: Covariate dataset shifts

Next, we will be looking into prior probability shifts, also known as label shifts. This is the opposite of a covariate shift, as this shift occurs when the output distribution changes for a given output but the input distribution remains the same.

In mathematical terms, this occurs only in Y -> X problems. When the prior probability changes, $p_{train}(y) \neq p_{test}(y)$, but the conditional probability remains the same, $p_{train}(x|y) = p_{test}(x|y)$

, a prior probability shift occurs:

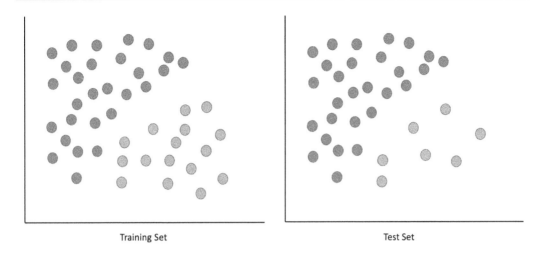

Training Set   Test Set

Figure 6.4: Prior probability shifts

Finally, we will discuss concept shifts, also known as concept drifts. This shift occurs when the distribution of the training data remains the same but the conditional distribution for the output given the training data changes.

In mathematical terms, this can occur both in X -> Y or Y -> X problems:

- For X -> Y problems, this occurs when the prior probability of the input variables remains the same in the training and testing datasets, $(p_{train}(x) = p_{test}(x))$, but the conditional probability changes, $(p_{train}(y|x) \neq p_{test}(y|x))$.

- For Y -> X problems, this occurs when the prior probability of the target variables remains the same in the training and testing datasets, $(p_{train}(y) = p_{test}(y))$, but the conditional probability changes, $(p_{train}(x|y) \neq p_{test}(x|y))$.

As an example, a user's purchasing behavior is affected due to the economy, but neither our training nor our test data contains any information regarding the economy's performance. As a result, our model's performance will degrade.

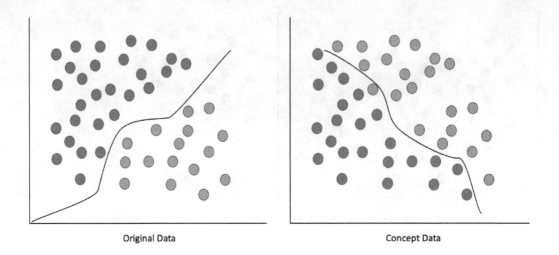

Figure 6.5: Concept shifts

This can be a tricky dataset shift since the distribution shift is not related to the data that we train on, but rather external information that our model may not have. Most of the time, these dataset shifts are cyclical and/or seasonal.

> **Important Note**
>
> Visualizing your data and calculating the different probabilities with regard to your data is the best way to help determine and identify which dataset shift you are dealing with. From there, you can decide how you will address your dataset shift.

When it comes to identifying most dataset shifts, there is a process that we can follow to help us. It includes the following steps:

- Preprocessing the data
- Creating random samples of your training and test sets on their own
- Combining the random samples into one dataset
- Create a model using one feature at a time while using the origin as the output value
- Predicting on the test set and calculating the **Area Under Curve – Receiver Operating Characteristics Curve (AUC-ROC)**
- If the AUC-ROC is greater than a certain threshold, for example, 80%, we can classify the data as having experienced a dataset shift

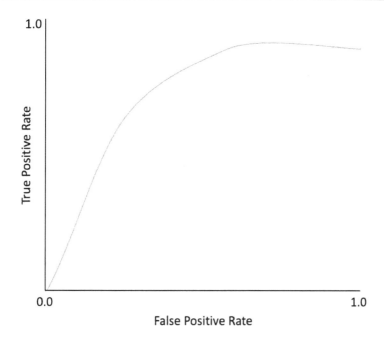

Figure 6.6: An example of an AUC-ROC graph (a value close to 1 indicates a strong model)

Handling and stabilizing dataset shifts

Now that we have established the methods for identifying the different types of dataset shifts, we can discuss the different ways of addressing these shifts and stabilizing our machine learning models.

While there are many ways to address dataset shifts, we will be looking at the three main methods. They consist of the following:

- **Feature dropping**
- **Adversarial search**
- **Density ratio estimation**

We will first look at feature dropping. This is the simplest form of adjusting dataset shifts. As we determine which features are classified as drifting, we can simply drop them from the machine learning model. We can also define a simple rule where any features with a drift value greater than a certain threshold, for example, 80%, can be dropped:

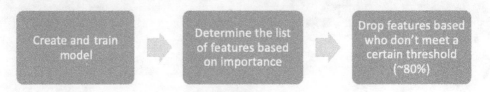

Figure 6.7: Feature Dropping Process

While this is a simple change, this is something that needs to be considered carefully. If this feature is considered important when training your machine learning model, then it is worth reconsidering whether this feature needs to be dropped. Also, if the majority of your features pass the threshold for being dropped, you may want to revisit your data as a whole and consider a different approach when addressing your dataset shift.

Next, we will look at adversarial search. This is a technique that requires training a binary classifier to predict whether the sample data is within the training or test datasets. We can then evaluate the performance of the classifier to determine whether there has been a dataset shift. If the performance of our classifier is close to that of a random guess (~50%), we can confidently determine that our training and test dataset distribution is consistent. On the other hand, if our classifier performs better than a random guess, then that will indicate an inconsistency between the distribution of the training and test datasets.

The adversarial search can be split into three parts:

1. From the original dataset, we will remove the target value column and replace it with a new column that indicates the source of data (train = 0 and test = 1).

2. We will create and train the new classifier with the new dataset. The output of the classifier is the probability that the sample data is part of the test dataset.

3. Finally, we can observe the results and measure the performance of our classifier. If our classifier performance is close to 50%, then this indicates that the model is unable to differentiate whether the data is coming from the training or test set. This can tell us that the data distribution between the training and test datasets is consistent. On the flip side, if our performance is close to 100%, then the model is confident enough to find the difference between the training and test datasets, which then indicates a major difference between the distribution of the training and test datasets.

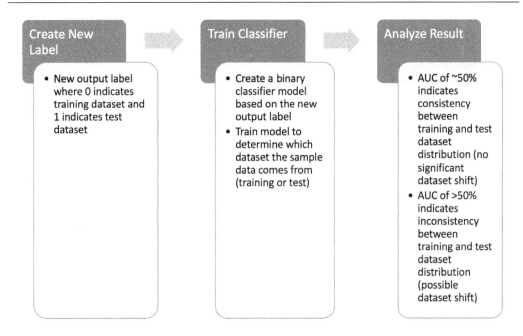

Figure 6.8: Adversarial search process

Using adversarial search, we can establish three methods to address the dataset shifts we encounter:

- Using the results, we can use them as sample weights for the training process. The weights correspond to the nature of how the data is distributed. The data that is similar in the actual distribution will be assigned a larger weight while that with inconsistent distribution will be given a lower weight. This will help the model emphasize the data that actually represents the real distribution it is trying to learn.

- We can use only the top-ranked adversarial validation results. Rather than mitigating the weights of inconsistent samples in the testing dataset, we can remove them altogether.

- All data is used for training except for the top-ranked adversarial validation results. This method can address the issues that can arise from the second method by using all the data rather than dropping features. Rather than discarding unimportant data, we can incorporate some of the data in the training data for each fold when using K-fold cross-validation during training. This helps maintain consistency while using all the data.

The final method used to address dataset shifts is called the density ratio estimation method. This method is still under research and not a commonly used method to address dataset shifts.

With this approach, we would first estimate the training and test dataset densities separately. Once we have done this, we will then estimate the importance of the dataset by taking the ratio of the estimated densities of the training and test datasets. Using this density ratio, we can use it as the weight for each data entry in our training dataset.

The reason this method is not preferred and is still under research is that it is computationally expensive, especially for higher dimensional datasets. Even then, the improvements it can bring to addressing dataset shifts are negligible and not worth the effort of pursuing this method.

> **Important Note**
> Feature dropping is the easiest and simplest way to address dataset shifts. Consider using this approach before using the adversarial search approach, as that option, while effective, can be a little involved and may require more effort and resources to help mitigate the effect of dataset shifts.

Summary

In this chapter, we went over the general concepts of dataset shifts and how they can negatively impact our machine learning model.

From there, we delved in deeper into what causes these dataset shifts to occur and what different characteristics dataset shifts can exhibit. Using these characteristics, we can better identify the type of dataset shift – whether it was a covariate shift, prior probability shift, or concept shift.

Once we were able to analyze our data and identify the type of dataset shift, we looked at different methods to help us handle and stabilize these dataset shifts so that we could maintain our machine learning model. We went over some techniques, such as feature searching, adversarial search, and density ratio estimation, that can assist us when dealing with dataset shifts.

Using these processes and methods, we can prevent our model from suffering from common dataset shifts that occur in the real world and continuously maintain our machine learning model.

Now that we have a firm understanding of machine learning and how to maintain a robust model, we can start looking into how we can incorporate our machine learning models into our **Microservices Architecture (MSA)**.

7

How Machine Learning and Deep Learning Help in MSA Enterprise Systems

In the previous chapters, we analyzed the different general concepts of artificial intelligence, machine learning, and deep learning, and how they can be used for certain applications and use cases. From there, we looked at how to create an end-to-end machine learning system pipeline and the advantages it brings when establishing a robust system. Finally, we examined the different ways our machine learning model can degrade over time through data shifts, and the different ways we can identify and address them.

Having a firm understanding of the basics of machine learning, we can now begin to explore the use cases of machine learning in our **Microservice Service Architecture** (**MSA**) enterprise. In this chapter, we will go over the different concepts we will be proposing when integrating machine learning in to an MSA enterprise system to establish an intelligent MSA.

Machine learning MSA enterprise system use cases

The space for adding machine learning to MSA enterprise systems is broad and can be open for many use cases. We can use machine learning for different types of problems that we can encounter in MSA, such as the following:

- **System Load Prediction**: This will determine when a service is experiencing higher than usual loads and trigger measures to prevent the system from degrading due to excessive server loads.

- **System Decay Prediction**: Similar to system load prediction, this will monitor the microservices and try to predict and determine anomalies in the MSA enterprise, allowing users to act and prevent certain issues from arising and negatively impacting the performance.

- **System Security**: In the age of cybersecurity, it is important to be able to protect your MSA system from targeted attacks. By studying the behavior of your MSA system, the model can predict and detect attacks that could be impacting your system.

- **System Resource Planning**: As your system grows and evolves, being able to properly allocate resources and adapt to your system needs is a critical part when establishing your MSA enterprise system. With machine learning, we can learn which services require more resources and how much we need to scale in order to allocate the required resources efficiently and effectively.

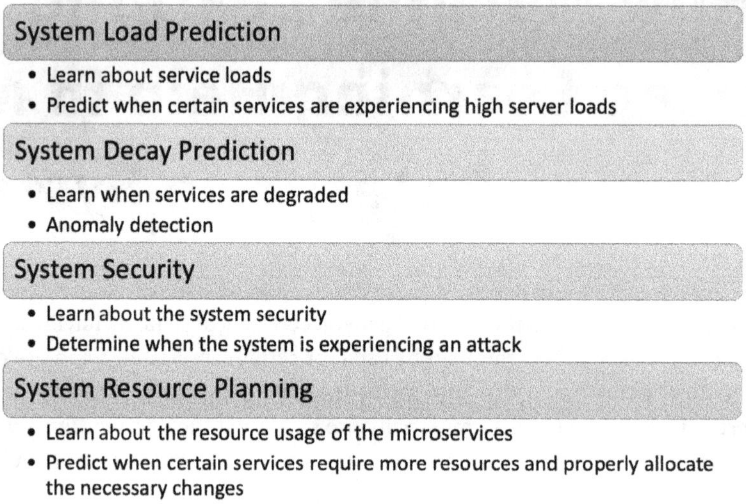

System Load Prediction
- Learn about service loads
- Predict when certain services are experiencing high server loads

System Decay Prediction
- Learn when services are degraded
- Anomaly detection

System Security
- Learn about the system security
- Determine when the system is experiencing an attack

System Resource Planning
- Learn about the resource usage of the microservices
- Predict when certain services require more resources and properly allocate the necessary changes

Figure 7.1: Use cases of machine learning in MSA

While there are many more use cases of machine learning in MSA enterprise systems, most use cases fall under these four categories. Before getting into the implementation of the different models, we need to first get an overview of the different cases and how we need to solve these different problems.

We can start by looking at system load predictions. This is a common issue that we will encounter when it comes to dealing with services in general. MSA has an advantage compared to monolithic systems, where the resources are dedicated to each microservice, allowing easier maintenance and scalability. As discussed in previous chapters, though, there could be cases where, in MSA, a microservice experiences a high load and, as a result, causes a cascading effect where the failures expand to other microservices.

With an intelligent MSA, we can train a model using different features, such as the response time, to learn the patterns of the MSA system. Similar to a microservice circuit breaker, this model will be able to swiftly determine whether a microservice is experiencing a heavy load and address the issue before it becomes too late and starts negatively impacting the other microservices.

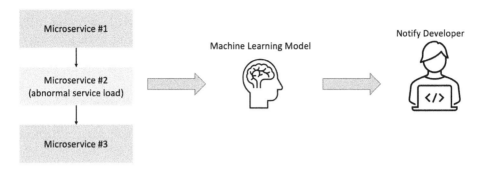

Figure 7.2: System load prediction model

Just like the system load prediction model, we can build a model to find anomalies within the MSA that could lead to decaying services. Rather than focusing only on the service load for a specific microservice, we can study the entire MSA and learn the different patterns of how it operates at a larger scale.

Certain systems can experience different system loads and bugs over certain times and periods. For example, our service may encounter spikes in requests over certain periods such as holidays and seasonal events, where the user count may drastically increase. Allowing the model to learn and understand the MSA and how it operates over time can prepare the model to better detect anomalies and prevent false positives.

Also, rather than monitoring separate microservices, we can evaluate clusters of microservices and how they interact with the entire MSA. This way, we can identify certain bottlenecks and bugs that could arise in our MSA.

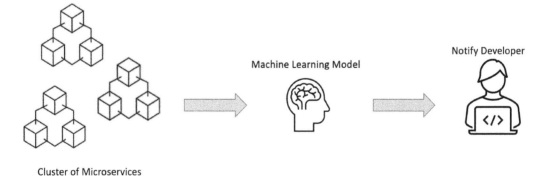

Figure 7.3: System decay prediction model

Machine learning has been thriving in the security field. With more advanced attacks and methods, it has become imperative for users to protect their systems. Machine learning has made it easier for users to create robust models that can analyze and predict attacks before they can even impact their systems, and MSA is no different.

Denial of Service (DoS) is a cyber-attack intended to prevent users from accessing certain services. These attacks are becoming more sophisticated with the advancements in technology. With machine learning, we can train our model to learn about our MSA and simulate DoS attacks such that it can be able to determine whether our MSA is under attack. With that, we can notify the security team or deploy countermeasures to fight back against certain attacks and maintain the integrity of our MSA.

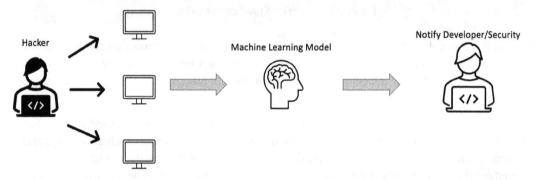

Figure 7.4: System security model

A part of the **self-healing** process includes resource allocation for certain microservices when your MSA begins to grow and expand. After a time, you may experience a growth in users and as a result, your microservices will have increased request volume. A model may incorrectly identify a problem and offer solutions that wouldn't address the core problem.

Thus, building an advanced model where it can track the gradual growth of the MSA and determine when certain services need more resources can be a critical part of the system's self-healing process. A successful implementation of the model can greatly improve system reliability as it can properly and efficiently allocate resources more effectively.

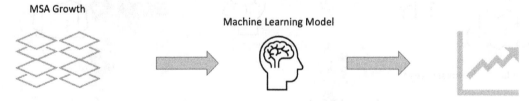

Figure 7.5: System resource planning model

> **Important note**
>
> The different types of models we can use in our MSA are not mutually exclusive. It is possible, and common, to combine the different use cases to build a more intelligent MSA. Understanding how your MSA operates and determining the different weaknesses it may have makes it easier for the user to determine which models to approach.

With certain use cases, some models can work better than others due to the nature of the problem. Now that we have looked at the different concepts where we can apply machine learning to our MSA, we can begin to dive deep into the different implementations and models we can use to build our machine learning models in the next few sections.

Enhancing system supportability and time-to-resolution (TTR) with pattern analysis machine learning

Before we can begin to make our MSA intelligent, we must first understand how our system performs by leveraging machine learning models to learn the common trends and patterns for the performance of our services. From there, we can establish a baseline that can be used as a reference for other advanced models to use.

As discussed in *Chapter 4*, supervised learning can occur when we have a labeled test set. For our case, we can mostly use supervised learning because we can easily capture the response time of our services in the MSA and use that as our data label.

From there, we have a wide variety of techniques that we can use to create our machine learning model. For simplicity, we can use a linear regression model to predict the expected response time for a particular microservice. Using this output, we can design a system where we can configure a set threshold where, if we detect that our MSA will reach a certain response time, we can notify the developers or initiate a program to resolve the issue before it occurs.

If we recall from *Chapter 6*, we discussed data shifts and how they can impact our model. It's common for MSAs to grow and expand as time passes due to an increase in user counts or seasonal occasions. As a result, we may see a growth in response times and metrics for our MSA. This may falsely trigger an alert notifying us of abnormal response times when, in reality, it accurately depicts the normal behavior of the MSA.

Therefore, it is important to continuously collect data and train our model to adapt to expected changes such that it is able to learn how the system grows and to correctly identify changes that are not common to our MSA.

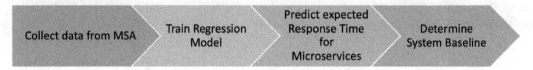

Figure 7.6: Performance baseline system flow

While this system is enough for simple problems, we can combine this model output with other advanced models to create a more end-to-end system, where we can understand the health of the MSA and make better decisions. In the next section, we will discuss how we can use deep learning to implement self-healing for our system.

> **Important note**
>
> It's important to start with a simple model, such as a linear regression model. Once the proof of concept works, you can improve your system by incorporating more advanced models and techniques.

Implementing system self-healing with deep learning

Now that we have determined the baseline for our system, we can use this to our advantage to create a more intelligent MSA, where we can detect anomalies and perform system self-healing. This way, we can be more proactive in resolving issues before they arise and save cost and time.

Anomaly detection is an effective method for identifying any abnormal events or trends that may occur in a system or service. For example, we can use anomaly detection for determining credit card fraud. We can use the user's purchasing trends and, based on that information, we can determine when the user has been a victim of credit card fraud.

Similar to credit card fraud detection, we can apply our anomaly detection to our MSA. Before we can go to the different models that we can use to achieve our anomaly detection, let us first understand the different types of anomalies:

- **Point Anomaly**: This occurs when an individual point is far off from the rest of the data
- **Contextual Anomaly**: Data is considered this way when it is not in line with the general data trend due to the context of the data
- **Collective Anomaly**: When a group of related data instances is anomalous with respect to the whole dataset

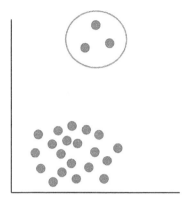

Figure 7.7: Anomalous data

An anomaly detection model can be done in the following ways:

- **Supervised Anomaly Detection**
- **Unsupervised Anomaly Detection**

A common model we can use for unsupervised learning is an **autoencoder**. As mentioned in *Chapter 4*, an autoencoder is a neural network composed of an encoder and a decoder. The general purpose of an autoencoder is to take the data and compress it to a lower dimension similar to PCA. That way, it is able to learn the correlations and patterns between the different data features. Once it learns the patterns, it can feed the compressed data forward to the decoder where it tries to "recreate" the original data with what it has learned in the encoder stage.

While experts can study the data to determine what response times are considered an anomaly for a particular MSA, we can leverage machine learning to help us find patterns and relationships that may be hard to see even for an experienced developer.

With the learned parameters, we can then use this in our supervised regression models to achieve more accurate results when detecting anomalies and prevent false positives from occurring.

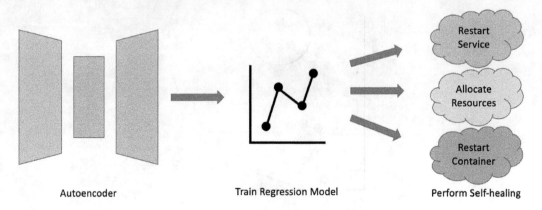

Figure 7.8: Self-healing using deep learning

> **Important note**
>
> Labeling data to be used in a supervised machine learning problem can cost time and money. You can leverage unsupervised machine learning models to help you predict and label your unlabeled data. From there, you can feed your newly labeled data into your supervised machine learning problem, thus taking advantage of unsupervised learning. Keep the newly labeled data in mind and make sure it doesn't negatively impact your supervised machine learning problem.

These are some of the ways in which we can take advantage of machine learning and deep learning to create an intelligent MSA, where it can detect anomalies in the system and react swiftly. These use cases can be adjusted and enhanced based on the user's needs and the demands of their MSA by using different models and techniques.

Summary

This chapter discussed how we can implement machine learning and deep learning in our MSA.

We first looked into the different use cases for how machine learning can be used to build an intelligent MSA. The uses cases can be grouped into four categories:

- System Load Prediction
- System Decay Prediction
- System Security
- System Resource Planning

We discussed each category and what role it plays when looking into creating an intelligent MSA.

From there, we started looking into using supervised machine learning to create a pattern analysis model where it can learn our MSA and create a performance baseline model. Using this, we can determine whether our microservice performance is abnormal. We can then use this to either perform actions based on a threshold or use this baseline to build a more advanced model.

Along with our supervised machine learning model, we can use deep learning to create a more sophisticated model, such as autoencoders, to find anomalies in our MSA. Using the combination of these two models, we can create a set of rules to perform based on certain predictions, such as that our MSA can self-heal with minimal human intervention. This allows us to save time and money when repairing and debugging our MSA.

In the next chapters, we will be taking what we've learned so far and starting to build our own MSA with practical examples and integrating machine learning to create our very own simple intelligent MSA.

Part 3:
Practical Guide to Deploying Machine Learning in MSA Systems

The final part of this book will bring everything covered so far to life. It will walk you step by step through the design and development of an intelligent **Microservices Architecture** (**MSA**) system, with hands-on examples and actual code that can be imported for real-life use cases. The part will provide an in-depth understanding of how to apply the DevOps process to building and running an intelligent enterprise MSA system, from the very start to operations and maintenance.

The part starts with the basics of containers, Docker, and how to install and run Docker containers. We will also gain hands-on experience in handling data flows between containers to build a simple project. Additionally, the chapter will cover a practical guide on building specific-purpose AI and how to infuse AI services into an MSA system.

This part delves into the application of DevOps to enterprise MSA systems, with a focus on organizational structure alignment and how DevOps can impact the MSA and its operations. We will learn how to apply DevOps throughout the project life cycle, from start to operations and change management and maintenance.

The part also covers how to identify and minimize system dependencies, apply **Quality Assurance** (**QA**) testing strategies, build microservice and MSA test cases, and deploy system changes and hot updates. The section will also provide practical examples of how to overcome system dependencies and apply testing strategies effectively.

In conclusion, the final part of this book will provide you with a comprehensive guide on how to design, develop, and maintain an intelligent enterprise MSA system, with a focus on practical, hands-on experience and real-life use cases. By the end of this part, we will be equipped with the skills and knowledge necessary to build our own intelligent MSA system and take the first step toward achieving better business results, operational performance, and business continuity.

This part comprises the following chapters:

8

The Role of DevOps in Building Intelligent MSA Enterprise Systems

In previous chapters, we covered what MSA is and the advantages of MSA over monolithic architecture. Then, we discussed, with examples, how to refactor a monolithic application into an MSA, and then talked about different patterns and techniques to enhance the performance of an MSA system.

We also discussed the different ML and DL algorithms with hands-on examples, how they can be optimized, and how these ML and DL algorithms can help further enhance the stability, resilience, and supportability of an MSA system in order to build a "smart MSA" or "intelligent MSA" system.

Over the next few chapters, we will further enhance our ABC-MSA system and try to apply what has been learned so far using some hands-on installations and code examples. However, before we do so, we need to discuss the different concepts of DevOps in this chapter, and how to apply the DevOps process to building and running an MSA system.

In *Chapter 1*, we briefly talked about DevOps in MSA. In this chapter, we will expand on the subject and dive into the details of the role of DevOps in building intelligent MSA.

The following topics are covered in this chapter:

- DevOps and organizational structure alignment
- DevOps processes in enterprise MSA system operations
- Applying DevOps from the beginning to operations and maintenance

DevOps and organizational structure alignment

In a traditional software development organization, the software delivery process is matured and built according to how that traditional organization is structured. Typically, we have a business team that defines the core business specifications and requirements, followed by another team of architects that builds how the system is supposed to be structured. In the traditional software model, we also have design engineers who write the functional specs, a development team responsible for writing the code, a QA team to test the code quality, then a release team, an operations team for post-release operations, a support team, and so on.

Figure 8.1: Traditional development structure

With all these teams involved in the pipeline in the traditional software release cycle, mostly sequential hand-offs between teams, silos, dependencies in between, cross-communication issues, and the possibility of finger-pointing during the process, the release cycle can take weeks or months to finish. For an MSA, this is not acceptable.

> **Important note**
> The whole purpose of MSA is to simplify, speed up, and optimize software releases and updates. Applying the traditional methodology to MSA system development just doesn't work and defeats the purpose of adopting an MSA to begin with.

DevOps

DevOps is one of the major processes adopted in modern software development organizations to help streamline the release process and optimize it so that an organization can make multiple seamless release updates every day with no service interruption whatsoever.

DevOps is a combination of processes that allow you to take an application from development to operation smoothly. Enterprises need dedicated and well-defined DevOps processes to manage their solution development, hosting, and operations.

The primary need of a DevOps team is to implement engineering techniques in managing the operations of applications. While this sounds simple to do, several mundane and random activities are carried out by the operations teams. Streamlining these tasks is the biggest challenge in adopting DevOps.

Figure 8.2: Teams working together in a DevOps fashion

The primary responsibility of the development team is to build the application. However, they also need to take care of other aspects of the application, such as the application performance, usage analytics, code quality, activity logging, and solving code-level errors.

On the other hand, the operations team faces a completely different set of problems. Their concerns include managing the availability of the applications, ensuring performance through higher scalability, and improving the monitoring of the solution ecosystem, the allocation of resources, and the overall system analytics. DevOps processes handle all of these concerns for all parties involved in the process.

Figure 8.3: DevOps life cycle

Figure 8.3 is similar to what we discussed in *Figure 1.11*. One new thing to add here is that the **PLAN** stage is where the software roadmap is defined and gets broken down into major requirements, called **epics**. These epics are broken down into a collection of short end user requirements, called **user stories**. More info on that will come in the next section.

Well, OK then, if an organization is to adopt an MSA, they should embrace a DevOps culture as well. Simple, right? Not quite!

Adopting a DevOps culture within a traditional organizational structure would have many misalignments that are guaranteed to hinder the DevOps cycle. The efficiency and speed of your release cycle will be as fast as the slowest process in your cycle. The software development organization itself has to shift its culture to align with DevOps, not the other way around. Many other methodologies and technologies will need to be adopted as part of the new shift to DevOps. The organizational structure itself may also need to be tweaked to align with the new DevOps methodologies.

The DevOps team structure

Setting up a DevOps team is the first step toward organizational transformation. However, you cannot expect to have a fully-fledged DevOps team without considering the existing organizational structure and how the organization is aligned with the existing development cycle.

It is imperative to have an interim phase in which the development and operations teams can function reasonably within the existing traditional organization. Both traditional Dev and Ops teams then slowly morph themselves into a true DevOps structure as the organization modernizes its structure to fit into the new culture.

One of the recommended approaches in the organizational transformation scenario is to develop a small DevOps team to work as a link between the existing development team and the operations team. The DevOps team's main objective in this particular case is to cross-function between both Dev and Ops teams to map deliverables in between, slowly familiarize both teams with the new methodology, and start applying basic DevOps methodologies within both teams so that they can be unified in the future.

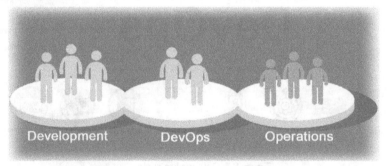

Figure 8.4: The DevOps team as a link between Dev and Ops during the organizational transition

Team communication, collaboration, energy, trust, and a solid understanding of the entire development cycle are all paramount to the new DevOps team's success. Therefore, you must identify the right skills and people who can push the activities of the DevOps team forward. These skills may include, but are not limited to, coding skills, mastering DevOps and **Continuous Integration/Continuous Development (CI/CD)** tools, and automation.

As the organizational structure and the teams mature and become more familiar with the new methodologies, merging the old Dev, old Ops, and the interim DevOps teams into a single new DevOps team becomes essential. Staying in the interim stage too long is likely to create even more disruptions than using the traditional development cycle for developing the MSA system.

The size of the DevOps team can be as small as 3 engineers, and as large as 12, depending on the organization's size, existing structure, and the effort being put into the organizational transformation. Usually, a number between 3 and 12 is ideal. Having a larger team is likely to create more challenges than benefits and start negatively impacting the team's overall performance.

Begin the process of transformation in a step-by-step manner, starting with infrastructure codification, the automation of infrastructure provisioning, source code version control, infrastructure monitoring, code build automation, deployment automation, test orchestration, cloud service management, and so on.

We know now how the organizational structure is relevant and important when embracing DevOps. We still need to understand some other details on the processes that will complement DevOps in order to achieve our goal of developing an efficient, high-quality MSA system with a short time-to-market and seamless updates.

In the following section, we will examine some other considerations that need to be taken into account when developing an MSA system.

DevOps processes in enterprise MSA system operations

Microservices development is a fast-paced process and requires all other development processes to run at the same pace. Right from the beginning of the development of the MSA system, source code management and configuration management are needed to provide the correct support to the DevOps team. This is followed by code scans and unit test orchestration in the development environment.

Having specific standard methodologies and best practices applied among the different team members is essential to manage the efficiency and fast pace of the development cycle. The following discusses what **the Agile methodology of development** is and how it helps in DevOps operations, and the importance of automation in DevOps.

The Agile methodology of development

Defining and accomplishing DevOps processes go hand in hand with adopting a development methodology that can fully support and leverage the power of DevOps. Although there are many ways to apply DevOps methodologies within your organization, the Agile methodology is the one best suited for DevOps.

The Agile development methodology breaks down the main requirements into small consumable changes – stories and epics. These small, consumable increments help the team achieve short wins throughout the journey of handling the project from start to end.

As shown in *Figure 8.5*, the Agile team members meet periodically, typically every week or two, to plan, define, and agree on the epics and stories. These requirements are then put into a backlog and, until the next Agile team meeting, the team members work to deliver the requirements from that backlog:

Figure 8.5: Sprint cycle in Agile development

In Agile development, the weekly or biweekly recurring meetings are called **Sprint Planning Meetings**, and the time between these meetings when developers are working on the backlog is called a **sprint**.

In order for team members to check on the status of each defined epic and story, they usually meet daily to examine the sprint backlog and refine whatever needs to be refined to ensure timely delivery. This daily meeting is called a **Daily Scrum**.

The Agile team handles continuously evolving user stories and requirements within a sprint cycle.

In an endeavor to deliver a high-quality product at a fast pace and low cost, Agile teams apply the following principles:

- No blocking time for day-end activities, such as building and deploying the latest code

- Immediate feedback on the code quality and functional quality of the latest code

- Strong control, precision monitoring, and continuous improvement of the daily activities of the development team

- Faster decision-making for accepting new stories, releasing developed stories, and mitigating risks

- A reduced feedback loop with the testers, end users, and customers

- Regular review and introspection of the development and delivery processes

A development team abiding by the Agile manifesto and following all the Agile principles should always look for ways to remove unwanted roadblocks from their process model.

The Agile methodology of development can be applied to develop and deliver all types of software projects; however, it is more suited to the development of microservices-based applications. It is important to view the scope and structure of microservices to align them with Agile and DevOps practices.

One of the most important pillars of the Agile and DevOps process is the use of on-demand, needs-based resources. This is usually catered to by the use of a cloud-based infrastructure. All the resources required by the Agile teams developing microservices need to be provisioned promptly and in the right quantity or with enough capacity. Cloud infrastructure is best suited to these requirements. Resources can be scaled up and down based on need and demand.

On-demand cloud workloads needed during the DevOps cycles are not necessarily deployed on the organization's private infrastructure; they may very well be deployed using a public cloud provider, or they may be deployed in a hybrid cloud fashion.

Automation

With the increase in the complexity of the IT infrastructure and MSA adoption and the demand for an Agile development cycle and short time-to-market, the need to streamline the infrastructure management processes becomes the most pressing need for any organization. A big part of managing an MSA's infrastructure, DevOps, CI/CD, and Agile development is automation.

Automation provides immense benefits to modern organizations. A few of these benefits include, but are not limited to, the following:

- **Better human resource utilization**: With automation in place, staff can focus on other activities that may not be automatable, hence optimizing the use of the organization's workforce, scaling better on other projects, and distributing responsibilities according to the available and required skill sets.

- **Better time-to-market and better business agility**: An automated process can certainly save a lot of time that would be otherwise consumed by manual repetitive work and potential dependencies. A job that may traditionally take days can be done in minutes when automation is in place.

- **Higher reliability and greater business continuity**: Complex and time-consuming tasks are simplified into simple keystrokes or mouse clicks. Accordingly, human error is significantly minimized, and operational reliability is largely increased.

- **Better compliance**: Compliance can be built into automation tools, providing better policy enforcement with minimum effort. Compliance includes industry compliance, best practices, and organizational standards as well. Industry standards may include the **General Data Protection Regulation (GDPR)**, **Payment Card Industry Data Security Standard (PCI DSS)**, **Health Insurance Portability and Accountability Act (HIPAA)**, and **Safeguard Computer Security Evaluation Matrix (SCSEM)**.

Automation is often used for the fast-paced and high-quality delivery of applications. DevOps is the key process that helps automate various phases of development and delivery. In fact, DevOps is the culture that helps organizations avoid repeated, time-consuming manual steps and efforts. There are various tools, frameworks, and processes within the ambit of DevOps that are needed for successful automation.

Most of the challenges within DevOps and MSA operations cannot be addressed manually – hence, the need for automation in DevOps and MSA is extremely high. Automation is needed in every area of delivery, from the time the microservice is developed to the time the microservice is deployed in the production environment.

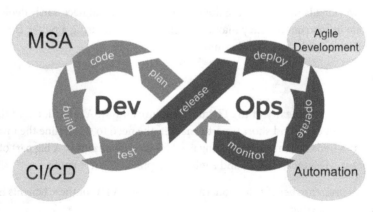

Figure 8.6: The four pillars of DevOps

In essence, modern enterprise system development needs DevOps to be able to respond to the dynamic and constantly growing needs of organizations, and DevOps depends heavily on four pillars: **MSA**, **Agile Development**, **CI/CD**, and **Automation**. These four pillars, as shown in the preceding diagram, play a significant part in DevOps success, and hence, in the success of modern enterprise system development.

Moreover, as we will discuss later in this chapter, AI applications are very hard to test and manage manually, and automation plays a big part in managing the entire DevOps cycle of AI applications.

Applying DevOps from the start to operations and maintenance

Every step of a microservices rollout requires a corresponding DevOps step. The confluence of the microservices development process with the DevOps process helps empower the Dev and Ops teams. The following is a detailed look at different facets of the DevOps process.

Source code version control

The Agile teams working on microservices require specific version control to be in place. Three aspects of version control need to be carefully defined for each microservice:

- The setup and management of version control tools, such as Git, SVN, CVS, and Mercurial.

- The version format and nomenclature for the application, such as a format to indicate the application version, the major-change version, the minor-change version, and the build or patch number – for example, version 2.3.11.7.

- The branching strategy for the source code. This is extremely important for microservices development with multiple teams working on separate microservices. Teams need to create separate repositories for each microservice and fork out different branches for each major or minor enhancement.

Configuration management and everything as a code

Configuration management is the practice of managing changes systematically across various environments so that the functional and technical performance of the system is at its best. This includes all the environments needed to develop, test, deploy, and run the MSA system components.

With so many moving parts in an MSA enterprise system, it is essential to identify which parts of the system need their configuration to be maintained and managed. Once these parts have been identified, their configuration will need to be controlled and regularly audited to maintain the overall health of the entire MSA system.

As the DevOps process matures, and as the MSA system components mature, things become very complex to manage and configure manually, and automation becomes critical for smooth and successful configuration management.

Configuration management tools can automatically and seamlessly manage the different aspects of the system components. These tools make adjustments as needed during runtime and whenever else, and in accordance with the version of the application, the type of change, and the system load.

One of the objectives of DevOps is to codify all the aspects of development as well as deployment, including the infrastructure and the configuration. The entire environment can be built from the ground up and quickly provisioned using **Infrastructure-as-a-Code (IaaC)** and **Configuration-as-a-Code (CaaC)**.

IaaC and CaaC are essential components of configuration management. Both are descriptive files typically written in languages such as Ansible, Terraform, Puppet, Chef, or CloudFormation.

With IaaC and CaaC, DevOps teams can easily spin up new workloads for different purposes. Workloads can, for example, be configured for testing, specify the properties of each workload based on the test cases involved, and control deviations from the main workload parameters.

CI/CD

As pointed out earlier in *Chapter 1*, CI/CD is an integral part of DevOps and plays the most important role in releasing MSA system updates. CI/CD ensures that the code is immediately and periodically built and pushed into the CI/CD pipeline for quick testing and feedback.

As shown in the following CI/CD pipeline diagram, developers focus primarily on working on the sprint backlog and push the code updates to the team repository, and it gets downloaded from there to the CI server.

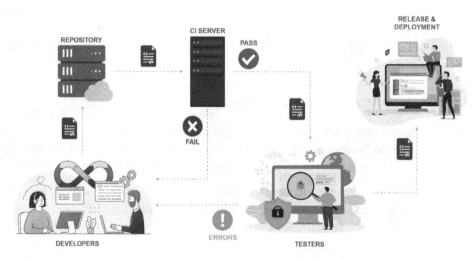

Figure 8.7: CI/CD pipeline and process flow

The CI server automatically runs preset test cases against the code and then pushes the code to the testers upon it passing all the test cases. Should any of the automated tests fail, the code doesn't move further along the pipeline, and an error report of all the test failures is sent back to the developers.

In contrast to the traditional development cycle, in which developers may find out about their code test results days or weeks after their code has been submitted for testing, in CI/CD, developers will get a report of their code problems within minutes. This early visibility into code errors gives developers the chance to immediately work on fixing these errors while working on the original code. Hence, they can continuously enhance the code for release and deployment.

Upon the code successfully passing all CI server tests, the code is tested further by the DevOps team testers. Testers then either push the code to release and deployment if no errors are found or return it for further fixes and enhancements.

This CI/CD pipeline enables developers to make frequent code merges; do unit testing, integration testing, code scans, and smoke testing; release; and deploy multiple times every single day – something that is not remotely possible using a traditional development cycle.

The DevOps team needs to identify a tool that can manage the entire CI/CD pipeline. DevOps helps add hooks and steps to include external executables and scripts for performing additional activities during the code build and deployment. Some of the most common and widely used CI/CD tools include Jenkins, Bamboo, and CircleCI.

Code quality assurance

Ensuring high-quality code, both in terms of coding standards and security vulnerabilities, is another important activity within DevOps. This is in addition to ensuring the accuracy of the application's business logic itself.

Code quality touches upon the concept of static and dynamic analysis of the code. Static analysis of the code is performed on the code itself before it gets executed. It is meant to uncover code smells, dirty code, vulnerable libraries, malicious openings in the code, and violations of code standards or best practices.

Dynamic code analysis is performed on the application during or after its execution. It is meant to uncover runtime errors due to the load, unexpected input, or unexpected runtime conditions in general.

Many tools that help perform code scans as part of CI/CD are available. These include, but are not limited to, SonarQube, Fortify SCA, and Raxis.

> **Important note**
> Testing AI applications is more challenging than testing regular applications. Certain aspects of AI applications do not exist in regular applications.

AI applications are non-deterministic – how they will behave in real situations is somewhat uncertain. Accordingly, expecting a specific outcome during AI application testing may not be viable. It may very well happen that the application being tested produces different outcomes with the same input or test criteria.

Most AI applications are as good as their training data quality, which makes AI applications subject to training data bias or unconscious bias. Imagine, for example, you are writing an AI module to predict home prices in any part of the United States, but your training data is 90% from a specific region within a specific state. Your AI model will accordingly be biased toward the area from which 90% of the training data came, so testing the AI application may require running tests against the training data itself. This may sound easy in this home price prediction case, but how would you make sense of other pieces of training data in more complex situations?

Let's assume that we can accurately test AI/DL applications despite all the training data challenges and their non-deterministic behavior. AI/DL applications constantly learn, train, and change their behavior, so by the time the code is running in production, the application is already learning and changing its behavior. The tests that have been completed a day or a couple of days earlier may not be valid anymore.

There are, of course, ways to overcome all these challenges. First of all, you will need to curate and validate the training data. You may need to perform both automated and manual tasks to validate the training data, including checking for data biases, data skews, distribution levels, and so on.

We will also need to test the AI algorithm and how the regression model performs against different sets of test data. The variance and mean square error of the model will also need to be examined and analyzed.

AI application testing tools are available on the market today and grow in number every day. The quality of these tools is constantly improving and can be a huge help to DevOps teams. AI testing tools are usually specialized based on the AI algorithms being used. Examples of different AI test tools include, but are not limited to, Applitools, Sauce Labs, and Testim.

Monitoring

With the advent of DevOps, standard monitoring has upgraded to continuous monitoring and covers the entire development cycle, from planning to deployment and operations.

Monitoring covers different aspects of the DevOps process and the components needed for the entire application to be developed, tested, deployed, and released, as well as for post-release operations to ensue. This includes infrastructure monitoring and the application itself.

Infrastructure monitoring includes the on-premises infrastructure, virtual cloud environments, networks, communications, and security. Application monitoring, on the other hand, involves performance, scalability, availability, and reliability. Resource monitoring includes the management and distribution of resources across multiple pod replicas within and beyond the physical or virtual workloads.

DevOps monitoring helps team members respond to any operational issues that arise during the DevOps pre-release or post-release cycles, hence enabling the DevOps team to be able to rectify, readjust, and make any necessary changes during the CI/CD pipeline.

Ideally, monitoring alerts trigger automatic actions to try to respond and fix a problem that has been detected. However, knowing that's not always possible, manual intervention is usually needed. Monitoring helps the DevOps team shift left to earlier stages in the development cycle to enhance their test cases, and accordingly, increase the application quality and minimize operational problems later on in the development cycle.

AI algorithms, as discussed earlier in *Chapter 7*, and as we will give more examples of later in this book, can detect any application behavior anomalies and automatically try to self-heal to prevent application operations from being disrupted.

There are many environment-specific tools available for DevOps monitoring, including Nagios, Prometheus, Splunk, Dynatrace, and AWS CloudWatch for AWS cloud environments.

Disaster management

Disaster management is an important yet often overlooked part of the DevOps process. In most cases, application recovery is seen as an extended part of the deployment process. In the cloud, it is generally considered to be an offshoot of configuring availability zones and regions for hosting an application instead of a full-fledged environment challenge.

In the case of microservices, identifying a disaster is a greater challenge than averting, mitigating, or managing it. Luckily, the CI/CD environment itself can be leveraged to test and simulate disaster scenarios. Moreover, the use of external repositories can be leveraged to recover code down to specific version numbers.

Nevertheless, setting up a completely separate set of environment replicas in different geographical locations, setting automatic failover, and load balancers in between can be great ways of maintaining business continuity and an uninterrupted CI/CD pipeline.

Using IaaC and CaaC tools to automate recovery is extremely helpful in bringing your applications and systems back online in minimal time in case of interruption.

You still need to define an incident response playbook as part of your DevOps. This playbook should include a detailed plan of what should be executed in each scenario. For example, a response to a natural disaster is likely different from a response to a data breach incident. The playbook needs to have different scenarios and a list of roles and procedures that need to be taken to prevent or minimize system interruptions or data loss.

Summary

For MSA systems to achieve the goals for which they were created, a certain set of methodologies will need to go hand in hand with developing an MSA system. In this chapter, we discussed a few of the most critical practices to embrace when developing an MSA system: the Agile methodology of development, DevOps processes and practices, and CI/CD pipeline management.

We also discussed how important it is to set up a DevOps team for managing microservices. We have given examples of tools to use to apply and manage DevOps when building our MSA system.

In the next chapter, we will take our first step in building an intelligent MSA system. We will talk about Docker, what it is, and why it's relevant. We will also create isolated and independent virtual environments using Docker and then link these environments (or containers) together to deliver a simple functional part of our MSA system.

## 9

Building an MSA with Docker Containers

In the previous chapter, we discussed how to apply the DevOps process of building and running MSA systems, and the importance of aligning the organizational structure with DevOps.

We also highlighted the importance of embracing automation and adapting agile development methodologies throughout the MSA project life cycle, and throughout the CI/CD operations.

This chapter will cover what a **container** is, how to install containers, how to work with them, and how to handle the data flow between containers to build a simple project. We will use **Docker** as our platform since it is one of the most popular and widely used platforms in the field today.

The following topics will be covered in this chapter:

- What are containers anyway, and why use them?
- Installing Docker
- Creating ABC-MSA containers
- ABC-MSA microservices inter-communication

What are containers anyway, and why use them?

A container is defined as an operating system-level virtualization artifact, created by grouping different finite compute resources into a self-contained execution environment.

As shown in the following figure of a container ship, containers are self-contained units, independent from any other container in the ship. The ship is the engine that is used to carry and transport containers:

Figure 9.1: A container ship

Similarly, the idea with containers is to create operating system-level virtualization. This means that, from within the kernel, you group different physical machine resources, applications, and I/O functions into a self-contained execution environment. Each of these self-contained resources forms a single container, hence the name container. The **Container Engine** is similar to the ship in the preceding example, where the container engine is used to carry, run, and transport the containers.

Containers have existed for a long time and can be traced back to Unix's **chroot** in the late 1970s and early 1980s, and before we even came to learn about what we call today a **hypervisor**. A hypervisor is a component that enables us to spin up **virtual machines (VMs)**.

Unix's chroot evolved later in the 1990s to Linux containers or what we call **LXC**, and then to **Solaris Zones** in the early 2000s. These concepts started to evolve with time from **cgroups** (originally developed by Google) and **namespaces** (developed by IBM) in to the container engines we see today, such as Docker, **Rkt**, **CRI-O**, **Containers**, Microsoft **Hyper-V** Containers, and more.

Although there are similarities between containers and VMs, both still have a few fundamental differences.

As shown in the following diagram, containers share the same kernel of the host operating system but isolate and limit the allocated resources, giving us something that feels like a VM but that's much more lightweight in terms of resources:

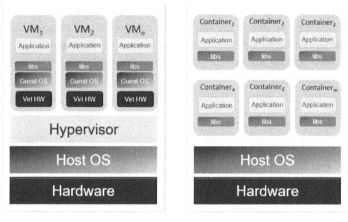

Figure 9.2: VMs versus containers

In hypervisor virtualization, each VM will have to have its own virtual hardware and its own guest operating system. In addition to all that, there is a great deal of emulation taking place in the hypervisor. Accordingly, each VM needs much more resources compared to what a container needs. Resources include CPU cycles, memory, storage, and more. Moreover, you are likely to have duplicates of the same guest OS deployed on multiple VMs for the VM to deliver the required function, thus even more overhead and waste of resources.

Figure 9.2 shows the hypervisor deployed on top of a host OS. A more common hypervisor virtualization model, however, is deploying the bare-metal hypervisor on the hardware directly. In either case, the overhead is are still significantly higher than deploying containers.

The lightweight nature of containers enables companies to run many more virtualized environments in the data center compared to VMs. Since containers share resources much more efficiently than VMs, and with finite physical resources in data centers, containers largely increase the capacity of the data center infrastructure, which means containers become a better choice in hosting applications, especially in our case of MSA.

Container performance is another thing to look at. With containers, I/O virtual drivers' communication, hardware emulation, and resources overhead are minimal, completely contrary to the case in the hypervisor virtualization environment. Accordingly, containers generally outperform VMs. Containers boot in 1-3 seconds compared to minutes in the case of VMs.

Applications running on a container can directly interact with and access the hardware. In the case of hypervisor virtualization, there is always a hypervisor between the application and the VM (unless a hypervisor bypass is enabled, which has its own limitations).

With containers, you can package the application with all of its dependencies in a contained environment that's reusable and completely portable to any other platform. That's one of the most important features of containers. Developers can develop an application on their development servers, ship it to the testing environment, then staging and production, and run the application without having to worry about portability and dependency issues.

> **Important note**
> For all the aforementioned reasons, the most popular deployment model of microservices is the container-per-service model. This is where each microservice of the MSA is deployed on a single container dedicated to running that particular application.

The other important difference between containers and VMs is security. Since containers use the same kernel, multiple containers may very well access the same kernel subsystem outside the boundaries of the container. This means a container could gain access to other containers running on the same physical host. A single application with root access, for example, could access any other container data.

There are many ways to harden the security of containers, but none of these techniques would help containers match the VM's total isolation security.

There are cases, of course, where using VMs would be a better option than using containers. Or in some scenarios, a mix of both VMs and containers would be the most appropriate deployment model. It all depends on the use case, the application, or the system you are deploying in your organization.

In a multi-tenant environment, where complete workload isolation is necessary, using VMs would be a better choice. Or, if you are trying to build an R&D environment for hosting critical intellectual capital, or highly confidential data or applications, a complete workload isolation will also be necessary. Therefore, in this case, using VMs would be the better option.

For our MSA example, we need a very lightweight, fast-starting, highly portable, and high-performing virtualization environment to build our MSA system. Hence, containers, with a container-per-microservice deployment model, are the better choice in our scenario. Each of the MSA system's microservices would be deployed in its own container and would have its own development team, development cycle, QA cycle, updates, run life cycle, and release cycle.

The following table summarizes the differences between containers and hypervisor virtualization:

	Containers	Hypervisor VM's
Resource Usage and Overhead	Lightweight, Please overhead, and more efficient use of resources	High overhead and resource-intensive
Container and Application Size	Averages 5-20 MB	Measured in 100s of MB or GB
Performance	High performance	Lower performance
Scalability	Easy to scale out/high horizontal scaling	Scaling out is harder and consumes resources
Bootup Time	Very short startup time (1-3 seconds)	Startup time is in minutes
Portability	System-agnostic and highly portable	Portability is limited
DevOps and CI/CD Suitability	Enables more agile DevOps and smoother CI/CD	Could slow down CI/CD operations
Host Hardware Access	Applications access HW directly	No direct access to HW
Security	Less secure; shares the same kernel	More secure; each VM has its own OS kernel

Table 9.1: Differences between containers and Hypervisor VMs

Despite the many options we currently have in choosing a container engine, Docker is by far the most popular engine used today, to the extent that Docker today is synonymous with containers. That's the main reason why we have chosen to work with Docker as our container's engine in this book.

Docker is also ideal for agile DevOps and CI/CD operations. In a CI/CD environment, the time between building a Docker image to the time it is up and running in the production environment is usually around 1-5 minutes in total:

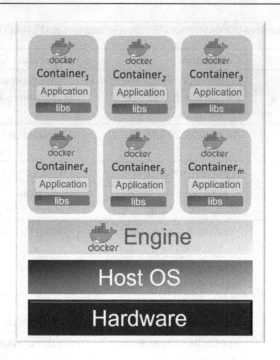

Figure 9.3: Docker Engine container virtualization

Figure 9.3 shows Docker Engine installed on the host operating system to enable the containerization of microservices or applications in general.

Docker in itself may not be sufficient to manage all the CI/CD operations. Organizations usually complement Docker by using a clustering technology such as **Kubernetes** or **Marathon** to smoothly deploy, manage, and operate the containers within the cluster in which your system is running. However, in this book, we will focus on Docker itself and how to use Docker to build our MSA system.

Also, to move, test, and deploy containers, we will need to have a repository to save these containers and be able to move them to different environments. Many tools can help with that, with **Docker Hub** and **GitHub** being two of the most commonly used repositories. For our project, we will use GitHub as our project repository.

So far, we have covered what containers are, the difference between containers and VMs, and why we prefer to use containers in MSA. In the next section, we will explain the different components of Docker, how to install Docker, and how to work with Docker's components to create a system's microservices.

Installing Docker

We will start this section by talking about Docker installation. Then, we will cover the main components of Docker, the purpose of each component, and how these components relate to each other. This section will help us prepare the environment that we will use later for our ABC-MSA demo project.

> **Important note**
>
> To maximize your hands-on learning experience, you need to follow all of our hands-on installation steps. But before doing so, please make sure you have a physical or virtual host available for the Docker installation demo before we dive deeper into this section. A virtual host can be created using virtualization software such as VirtualBox or VMWare.

Although you can install Docker on Windows or Mac, in our demo, we will use an Ubuntu Server 22.x Linux environment to install Docker **Community Edition** (**CE**). We suggest you use a similar environment to be able to follow our installation steps.

Docker Engine installation

Now that we know the main components of Docker, let's take a step back and learn how to install Docker and create different Docker images for the ABC-MSA system.

The best way to install Docker Engine is to follow Docker's official installation guide from Docker Docs at `https://docs.docker.com/engine/install/`. Pick your server system platform installation guide from the list.

You may also want to install Docker Desktop on your workstation. Docker Desktop is available for download from the same installation guide referred to previously.

After the installation is completed, verify Docker's functionality by running the following command:

```
$ docker --version
Docker version 20.10.18, build b40c2f6
$
```

And,

```
$ docker run hello-world

Hello from Docker!
This message shows that your installation appears to be working
correctly.
:

:
```

You may need root privileges to issue the Docker commands successfully.

Now that we have installed Docker, let's go over the main components of Docker and how to use each.

Docker components

There are four main components of Docker: the Docker file, the Docker image, the Docker container, and the Docker volume. The following is a brief description of each of these components.

The Docker file

The Docker file is a text file that works as a manifest that describes how the Docker image should be built. The Docker file specifies the base image that will be used to create the Docker image. So, for example, if you were to use the latest Ubuntu version as your base Linux image for the container, you would have the following line specified at the top of your Docker file:

```
FROM ubuntu
```

Notice that ubuntu is not tagged with any version number, which will instruct Docker to pull the latest version available for that base image. If you prefer to use CentOS version 7.0, for example, you must then tag the base image with the version number, as shown in the following line:

```
FROM centos:7
```

The specific image tag can be found on Docker Hub. Docker Hub is a public repository that stores many free Docker official images for reuse by Docker users. Among many others, base images could be Linux, Windows, Node.js, Redis, Postgres, or other relational DB images.

After you specify the base operating system image, you can use the RUN command to run the commands that you would like to execute during the Docker image creation. These are regular shell commands that are usually issued to download and install packages and libraries that will be used in your Docker image.

The Docker file has to be named Dockerfile for Docker to be able to use it. The following is a simple Dockerfile example:

```
FROM ubuntu:22.10

# Install some packages
RUN echo "Installing packages..."
RUN apt-get update && \
    apt-get install -y python3 python3-pip; \
    python3 -m pip install ansible
RUN echo "Installing packages complete!"
```

Figure 9.4: A Docker file (Dockerfile) example

The preceding sample Dockerfile does the following:

1. Uses Ubuntu version 22.10 as the base image to run on the container that will be created later.

2. Fetches the latest packages list.

3. Installs Python version 3 and the PIP Python package management system.

4. Installs a package called Ansible (Ansible is an automation tool).

The Docker image

Once you have finished composing your Dockerfile, you will need to save it as a `Dockerfile` to be able to use it to create the Docker image.

A Docker image is a binary file that works as a template with a set of instructions on how a Docker container should be created.

Please note that a Docker image can either be created from the Dockerfile, as we are explaining here, or downloaded from a public or private repository such as Docker Hub or GitHub.

To build a Docker image, use the following command while pointing at the `Dockerfile` location. The following example assumes the Dockerfile is located in the user's home directory:

```
$ docker build -t packt_demo_image ~/
```

The preceding command will build an image called `packt_demo_image`. This image will be used later to create the container with the specs defined in the Dockerfile.

The `-t` option means `tty`, which attaches a terminal to the container.

To verify that your image has been created, use the following command:

```
$ docker image ls
```

You can add the `-a` option to the end of the proceeding command to show all images created on the host machine.

In CI/CD operations, the images that are built are usually shared in a public or private repository so that they're available to the project team, or even the public in some cases.

The Docker container

The last step is to run a container based on the Docker image you created (or pulled from the image repository). To run a container, use the following command:

```
$ docker run packt_demo_image
```

To verify that the container is running, use the following command:

```
$ docker container ls
```

The preceding command will show only the running containers. To show other containers on the host machine, add the -a option to the end of the command.

You can also use the older version of the preceding command to verify that the container is running:

```
$ docker ps
```

The following diagram shows the relationship between all four Docker components and summarizes the entire process of running a container. First, we create a Dockerfile. Then, we use that file to create the Docker image. The Docker image can then be used to create the Docker container(s) locally, or first uploaded to a private or public repository where others can download and create their Docker container(s):

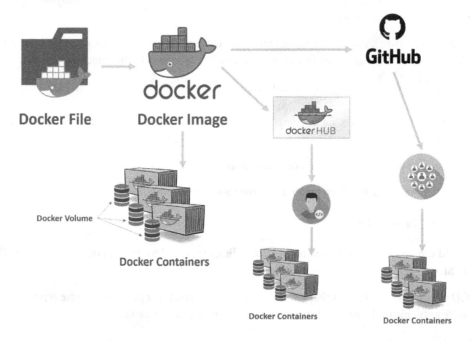

Figure 9.5: Docker components

Docker containers have a life cycle of their own – they can run for a specific task with no regard for what their previous state is, and once that specific task is completed, the Docker container automatically terminates.

In other cases, containers need to be aware of their previous status. If so, they will need to be persistent to preserve the container data after its termination. That's when Docker volumes become very handy. Next, we will talk about what a Docker volume is and how it can be created.

The Docker volume

Docker volumes are a form of storage that a Docker container can be attached to. Containers are attached to volumes to read and write persistent data, which are necessary for the function of the container.

To elaborate more, consider the Docker container for the Customer Management microservice (`customer_management`). If you need to create a new customer in the `customer_management` container, you will need to update the local data store installed in that container. If the container is not persistent, once the container terminates, all data created or changed inside that container will be lost.

To avoid this problem, we will need to create a Docker volume and attach the container to that volume. The container itself can then run and update whatever data it needs to update in its volume, and then terminate. When it starts the next time, it gets instantiated with all the previous statuses and data it had before the last termination.

To create a Docker volume for the `customer_management` container, for example, use the following command:

```
$ docker volume create customer_management_volume
```

The following command will list all volumes created on our host machine and verify the volume we have just created:

```
$ docker volume ls
```

Once we create the volume, Docker mounts a local drive space on the host machine to preserve the container's data and its mounted filesystem.

To show more details about the volume, including the volume's name, the local host and the container's target mount locations, and the date and time of the volume's creation, use the `docker volume inspect` or `docker inspect` command, as follows:

```
$ docker volume inspect customer_management_volume
[
    {
        "CreatedAt": "2022-10-14T22:24:46Z",
        "Driver": "local",
        "Labels": {},
        "Mountpoint": "/var/lib/docker/volumes/customer_
management_volume/_data",
```

```
        "Name": "customer_management_volume",
        "Options": {},
        "Scope": "local"
    }
]
```

Assuming we have previously created the `packt_demo_image` image, to create the persistent `customer_management` container, we will need to attach the container to the volume we have just created using the mount points shown in the `docker volume inspect` command's output. The following command will create the container, attach the volume to the container, and then run the container:

```
$ docker run -itd --mount source=customer_management_
volume,target=/app_data --name customer_management_container
packt_demo_image
```

The `it` option in the `docker run` command is for interactive `tty` mode, and the d option is for running the container in the background.

`/app_data` is an absolute path within the container that's mounted to the local host's mount point. From the preceding inspect data shown, the `/var/lib/docker/volumes/customer_management_volume/_data` mount point is mapped to `/app_data` in the container.

To verify that the container is running, use the following command:

```
$ docker container ls
```

If the container terminates for whatever reason, use the `-a` option at the end of the preceding command to show the available container on the host. You can use the `docker container start` or `docker container stop` command, followed by the container's name, to run or terminate any of the available containers you built on that host.

Now that we have installed Docker Engine and understand the different components of Docker, we will go over how to create the main ABC-MSA containers as microservices and provide an example of how these microservices talk to each other.

Creating ABC-MSA containers

In our ABC-MSA system, we are adopting a container-per-microservice approach. Therefore, we need to identify the main containers we will build, the components we need for each container in our ABC-MSA system, and then build the necessary Dockerfile(s) to use.

We are building our microservice applications using **Flask**. Flask is a **Web Server Gateway Interface (WSGI)** micro-framework that enables applications to respond to API calls in a simple, flexible, and scalable manner. We won't discuss our applications' code in this book, but the code is available on our GitHub with detailed documentation for your reference.

In this section, we will explain how we build our ABC-MSA Dockerfile(s), images, and microservices, how we will start to listen to API calls in each container, and how the system's microservices will be able to communicate with each other.

For demo purposes, we will use port HTTP/8080 in the container to listen to HTTP API requests. The production environment should use HTTPS/443 and consider the `tomcat` server for handling all web connections.

The following is only part of the full system container setup. All the ABC-MSA system's created files and Docker images can be found in our GitHub repository at `https://github.com/PacktPublishing/Machine-Learning-in-Microservices`.

ABC-MSA containers

The following are the services we have previously identified for our ABC-MSA system:

1. API Gateway
2. A frontend web dashboard interface
3. Customer Management
4. Product Management
5. Order Management
6. Inventory Management
7. Courier Management
8. Shipping Management
9. Payment Authorization
10. Notification Management
11. Aggregator: "Product Ordered Qty"
12. Management and Orchestration

We can code and build each of these services from scratch, but the good news is that we don't have to. Docker Hub offers a rich library with many Docker images that we can leverage in building our microservices. Docker Hub can be accessed at `https://hub.docker.com/`.

We will not go over each of these services. Instead, we will focus on the ones that provide different development and deployment approaches. Some of the services are already available through Docker Hub, and some others are similar, so one example of these will suffice. Nevertheless, all the project files will be made available in this book's GitHub repository.

API Gateway

Many open source and commercial API gateways can be pulled from different internet repositories, including Tyk, API Umbrella, WSO2, Apiman, Kong, and Fusio, to name a few. We will use Tyk in our ABC-MSA system since it is easy to use, has comprehensive features including authentication and service discovery, and is 100% an open source product with no feature restrictions.

To install a Tyk Docker container, just follow the instructions at `https://tyk.io/docs/tyk-oss/ce-docker/`.

By default, the Tyk API gateway listens to TCP port `8080`. To verify your installation, issue an API call test to Tyk using the `curl` command, as follows:

```
$ curl localhost:8080/hello
{"status":"pass","version":"4.1.0","description":"Tyk GW"}
```

If Tyk has been successfully installed and is running on your host, you should get a dictionary output stating Tyk's status and the current version, as shown in the preceding command output.

You can also verify that the Tyk Docker image and container were created successfully using the following commands:

```
$ docker images
REPOSITORY                               TAG
IMAGE ID        CREATED         SIZE
redis                                    6.2.7-alpine
48822f443672    3 days ago      25.5MB
docker.tyk.io/tyk-gateway/tyk-gateway    v4.1.0
0c21a95236de    8 weeks ago     341MB
hello-world                              latest
feb5d9fea6a5    12 months ago   13.3kB
$
$ docker container ls
CONTAINER ID    IMAGE
          COMMAND
              CREATED           STATUS          PORTS
                                NAMES
ac3ac1802647    docker.tyk.io/tyk-gateway/tyk-gateway:v4.1.0
```

```
   "/opt/tyk-gateway/ty…"    54 minutes ago    Up 54 minutes
     0.0.0.0:8080->8080/tcp, :::8080->8080/tcp
tyk-gateway-docker_tyk-gateway_1
9e0f1ecfb148    redis:6.2.7-alpine
"docker-entrypoint.s…"    54 minutes ago    Up 54 minutes
0.0.0.0:6379->6379/tcp, :::6379->6379/tcp
tyk-gateway-docker_tyk-redis_1
```

We can see the `tyk` image details in the preceding command output, as well as the running container and what port it is listening to. We can also see a Redis image and container. This is because Redis is a prerequisite for Tyk and is included in the Tyk installation package.

The Customer Management microservice as an example

The Customer Management, Product Management, Order Management, Inventory Management, Courier Management, Shipping Management, Payment Authorization, and Notification Management microservices are all similar in terms of how we can build and deploy the container. In this section, we will learn how to create an image that we can use to create a system microservice. We have picked the Customer Management microservice as an example.

As mentioned earlier, for these microservices to communicate with the API gateway or any other components in the ABC-MSA system, we need to have Flask installed and running, listening to port HTTP/8080 in the running container.

We also need an internal data store for our application to use and manage. And since our code will be written in Python, we need to have Python installed as well. All these required components, along with some essential dependency packages, need to be specified in our Dockerfile.

Now, we need to write the Dockerfile required for creating the microservice image that we will use to create the microservice container. Each ABC-MSA container should have its own development cycle and be deployed either using the CI/CD cycle we discussed in *Chapter 8* or uploaded manually to the team repository.

The following is an example of the Dockerfile that's required for creating the Customer Management image:

```
# Docker File for "customer_management" microservice
FROM ubuntu

# Install some dependencies/packages
RUN apt-get install -y apt-transport-https
RUN apt-get update
RUN apt-get install -y net-tools mysql-server python3 pip git
build-essential curl wget vim software-properties-common;
```

```
# Install OpenJDK
RUN apt-get update && \
    apt-get install -y default-jdk
ENV JAVA_HOME /usr/lib/jvm/java-11-openjdk-amd64/


# Install Flask to run our application and respond to API calls
RUN pip install -U flask


# Expose port TCP/8080 to listen the container's application/
flask API calls
EXPOSE 8080


# Create the /app_data directory and make it the working
directory in the container
RUN mkdir /app_data
WORKDIR /app_data
ENV PATH $PATH:/app_data


# Download the microservice app code from GitHub repo
ENV GIT_DISCOVERY_ACROSS_FILESYSTEM 1
RUN git config --global init.defaultBranch main
RUN git init
RUN git remote add origin https://github.com/mohameosam/abc_
msa.git
RUN git config core.sparseCheckout true
RUN echo "/microservices/customer_management/" > /app_data/.
git/info/sparse-checkout
RUN git pull origin main


# Initialize the flask app
ENV FLASK_APP /app_data/microservices/customer_management/
customer_management_ms.py


# Specify a mount point in the container
VOLUME /app_data


# Start mysql & flask services and available bash sheel
```

```
RUN chmod +x /app_data/microservices/customer_management/start_
services
CMD /app_data/microservices/customer_management/start_services
&& bash
```

The aforementioned Dockerfile specifies what the Customer Management Docker image should look like. The following are some insights into what each of the lines in the file will do:

1. Specify Ubuntu as the Linux operating system that will be used in the Customer Management container.

2. Install some required packages:

    -   MySQL (required for our application)

    -   Python (required for our application)

    -   pip (required to be able to install Flask)

    -   The rest are some other tools needed for troubleshooting (optional)

3. Install Flask (required for our application).

4. Expose TCP/HTTP port 8080 for Flask to listen to API calls.

5. Create a working directory in the container to act as the mount point for saving the container's data.

6. Download the Customer Management application code from our GitHub repository.

7. Set an environment variable to let Flask know what application it will use when responding to API calls.

8. Use our downloaded start_services shell script to start Flask and MySQL in the container.

The start_services shell script contains the following commands:

```
flask run -h 0.0.0.0 -p 8080 &
usermod -d /var/lib/mysql/ mysql
service mysql start
```

The first line enables Flask to listen to port 8080 on all the host network interfaces. This is OK in the development and testing environment. In the production environment, however, Flask should only be available on the localhost 127.0.0.1 network interface to limit API access to the local environment. Also, for better security, port HTTPS/443 should be used in API calls instead.

Assuming the Dockerfile has been placed in the current user home directory, we now need to create our Customer Management microservice/container from the Dockerfile:

```
$ docker build -t abc_msa_customer_management ~/
```

Docker will take a few minutes to finish creating the image. Once all the Dockerfile steps have been completed, you should see the following command as the last line of the `docker build` command's output:

```
Successfully tagged abc_msa_customer_management:latest
```

This signals a successful completion. Now, we can use the `docker image ls` command to verify that the `abc_msa_customer_management` image has been created successfully.

The last step is creating the container. Since the application will configure and update the MySQL database, we need to create a persistent container to retain all the changes.

Similar to what we explained earlier, we will use the `docker run` command to create the Customer Management container, as follows:

```
$ docker run -itd -p 8003:8080 --mount source=customer_
management_volume,target=/app_data --name customer_management_
container abc_msa_customer_management
```

The p option is used to "publish" and map the ports that the container listens to with the ports the host machine listens to. So, the host machine will be listening to port 8003 for HTTP/8080 requests on the container.

We have chosen 8003 to standardize the way the host listens to the container's API call requests.

Remember that each container has a TCP stack that is different from the host's TCP stack. So, the TCP HTTP/8080 port is only local within the container itself, but outside that particular container's environment, that TCP HTTP/8080 port is different from the TCP HTTP/8080 port available on any other container or on the host machine itself.

To access that port from outside the realm of the `customer_management` container, you need to map the `customer_management` container's TCP HTTP/8080 port to a specific port on the host machine.

Since we need to map the local TCP HTTP/8080 port of each of the 12 containers we identified earlier, we decided to follow a specific pattern. Map the TCP/80nn port on the host machine to each local TCP HTTP/8080 of each container. Here, nn is the container's number.

Figure 9.6 shows how some of the ABC-MSA container's TCP HTTP/8080 ports are mapped on the host machine.

We don't have to run all the containers on a single host. The system containers could be scattered across different hosts, depending on many factors, such as how critical the service/application running on the container is, how the system is designed, the desired overall redundancy, and so on:

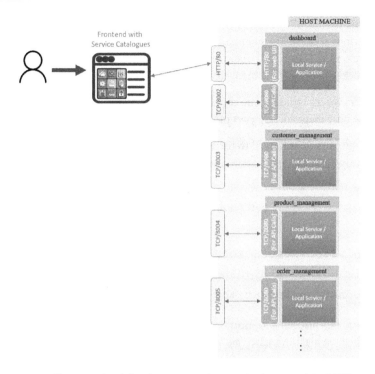

Figure 9.6: The container's local port mappings to the host machine's TCP stack

Now, verify that the container is running using the following command:

```
$ docker container ls
```

The following command will allow you to connect to the container's bash shell using the root privilege (a user ID of 0, as specified in the command):

```
$ docker exec -u 0 -it customer_management_container bash
```

That's all for the Customer Management microservice. In the same manner, we can create the rest of the ABC-MSA containers. We just need to make sure we use appropriate corresponding names for the other microservice's containers and volumes and map to the right TCP/80nn port number on the host machine.

The frontend web dashboard interface

The dashboard is the main component of the **user interface** (**UI**) interaction and interacts with all services offered to the user. In our ABC-MSA example, we created a simple cart application where the user can place products in the cart and place an order.

The Dashboard container is built the same way the `customer_management` container is built, as shown in the previous section. The main difference between both is the additional web server that we will need to have on the Dashboard microservice, and the ports to be exposed on the container. The Dashboard's Dockerfile should be changed accordingly.

Like all the containers we are building, the container's local TCP port that listens to API calls is TCP HTTP/8080, and the host-mapped TCP port in the `dashboard` container case should be TCP/8002.

The Dashboard container will still need to listen to HTTP/80 for user web UI requests. Unless the host machine is running another application or web page on HTTP/80 port, we should be OK to use that port.

Now, we need to map the HTTP/80 port on the host machine, as shown in the following `docker run` command:

```
$ docker run -itd -p 8002:8080 -p 80:80 \
--mount source=dashboard_volume,target=/app_data \
--name dashboard_container abc_msa_dashboard
```

This command has an additional p option to map the HTTP/80 port on the container with the HTTP/80 port on the host machine. `abc_msa_dashboard` is the Dashboard microservice image.

Managing your system's containers

As you saw in the preceding examples, the `docker run` command can get lengthy and messy. **Docker Compose** helps us manage the deployment of containers. With Docker Compose, it is much easier to manage the deployment of the containers, change deployment parameters, include all system containers in a single YAML file, and specify the order of the containers' deployment and dependencies.

The following is a sample YAML file for initializing three of the ABC-MSA containers, as we did with the `docker run` commands earlier, but in a more organized and structured YAML way:

```
# Docker Compose File abc_msa.yaml
version: "3.9"
services:

  customer_management_container:
    image: abc_msa_customer_management
    ports:
      - "8003:8080"
    volumes:
      - customer_management_volume:/app_data
```

```
product_management_container:
  image: abc_msa_product_management
  ports:
    - "8004:8080"
  volumes:
    - product_management_volume:/app_data

dashboard:
  image: abc_msa_dashboard
  ports:
    - "8002:8080"
    - "80:80"
  volumes:
    - dashboard_volume:/app_data
  depends_on:
    - customer_management_container
    - product_management_container

volumes:
  customer_management_volume:
  product_management_volume:
  dashboard_volume:
```

The following command runs the Docker Compose .yaml file:

```
$ docker-compose -f abc_msa.yaml up &
```

The f option is used to specify the YAML file's name, and the & option is used to run the containers in the shell's background.

In this section, we showed you how to create some of the ABC-MSA images and containers. The ABC-MSA containers are now ready to communicate with each other either directly or, as we will show later in this book, through the API Gateway.

In the next section, we will learn how we can use the containers we created, how we can issue API calls to them, and what response we should expect.

ABC-MSA microservice inter-communication

In this section, we will learn how to expose APIs from containers and how containers communicate with API consumers.

The microservice application code for each container is available in the ABC-MSA project on GitHub. We recommend that you download the code to your local test environment to be able to get some hands-on experience when following the steps we will cover in this section.

There are two main ways for containers to communicate with each other. One is by using the container's name in a **Docker network**, and the other is by using the container's IP and TCP port. The following are some of the details you need to know about to be able to configure your containers to communicate with each other.

The Docker network

When we have containers running on the same host, containers can communicate with each other on the same host using only container names and without the need to specify the container's IP address or listening port.

The concept of using only container names is programmatically very useful, especially in cases where these IPs change dynamically. The names are usually deterministic, and by only specifying the Docker's container name, you avoid having to apply different layers of system operations to first learn about the container's TCP/IP details before starting to communicate with the target container.

However, there are some prerequisites to enabling container communication through their names only:

- The containers communicating with each other will all need to be on the same host
- We will need to create a Docker network on the host
- We will need to attach the containers to the created Docker network when running the container using the docker run command or by specifying the container's instantiation details in the docker-compose YAML file

The following command creates a Docker network on the host machine that can be used for our ABC-MSA system's inter-microservice communication:

```
$ docker network create abc_msa_network
```

The following command lists the Docker networks configured on the host machine:

```
$ docker network ls
```

Now, attach the ABC-MSA containers to the `abc_msa_network` network by using the `--network` option in the `docker run` command, as shown in the following example:

```
$ docker run -itd -p 8003:8080 \
--network abc_msa_network --mount \
source=customer_management_volume,target=/app_data \
--name customer_management_container abc_msa_customer_
management
```

Using Docker networks is very useful in many cases. However, since we are designing our ABC-MSA system so that containers can run independently of their host location, we will be using the container's IP/TCP communication.

In the next section, we'll explain how the ABC-MSA microservices communicate using TCP/IP and go over some examples of how to test the communication and data exchanges.

TCP/IP communication between containers/microservices

So far, we have installed Docker, built our Docker images and volumes, and started the containers of all our microservices. Now, it is time to understand how these containers interact with each other.

Upon running Docker on the container's host, the host automatically creates a virtual IP network and assigns an IP address to each running Docker container on that host. That virtual IP network is only internal to the host running the containers and cannot be accessed from anywhere outside that host.

The container's host carries at least two IPs. There's one inside IP that's internal to the Docker network and that can only be recognized inside that Docker network. Then, there's an outside IP, which is usually assigned by the **Dynamic Host Configuration Protocol (DHCP)** server in the organization's network. The outside IP is necessary for the container's host to communicate with the outside world.

The internal Docker network of the container's host is not visible to any other host in the network. Therefore, for an outside host to communicate with a specific container in the container's host machine, it will need to use the outside IP of the container's host machine.

Among a lot of other information, to get the assigned IP address, as well as the inside and outside listening ports of a specific container in your system, use the `docker inspect` command, followed by the container's name.

Our demo setup is shown in the following diagram:

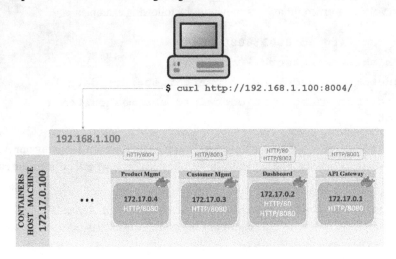

Figure 9.7: ABC-MSA container communication

As you can see, the host machine's inside IP address is 172.17.0.100, and the outside IP address is 192.168.1.100. The container's host is listening to the container's mapped ports (8001 to 8012), as explained earlier.

If other hosts in the network want to send API calls to one of the ABC-MSA containers, that outside host will need to send the request to the outside IP address of the container's host, 192.168.1.100, using the mapped port of the container it wants to communicate with.

To elaborate further, the preceding diagram and the following example show an outside host testing the API response of the Product Management container:

```
$ curl http://192.168.1.100:8004/
<!DOCTYPE html>
<head>
    <title>PRODUCT MANAGEMENT Microservice</title>
</head>
<body>
    <h3>Product Management Microservice Part of ABC-MSA System</
h3>
</body>
```

ABC-MSA API calls return a JSON variable for easier data handling. One of the APIs we built for ABC-MSA microservices is `service_info`. An example of an API call for `service_info` is as follows:

```
$ curl http://192.168.1.100:8004/api?func=service_info
{"service_name": "product_management", "service_descr": "ABC-
MSA Product Management"}
```

If you are communicating internally from within the Docker network (172.17.0.0), you can communicate directly with the container's IP and listening ports. Performing the same `curl` test on the Product Management container from the API Gateway shell would look like this:

```
$ curl http://172.17.0.4:8080/api?func=service_info
{"service_name": "product_management", "service_descr": "ABC-
MSA Product Management"}
```

Knowing how to pass API requests and handle the API response is key to developing your MSA system. Please refer to our ABC-MSA code in this book's GitHub repository for examples of how the API calls are issued and handled across the entire system.

Summary

In this chapter, we covered the concept of containers, what they are, and how they are different from VMs. Then, we worked with Docker as one of the most popular container platforms available today. We showed you how to install Docker and create Dockerfiles, Docker images, Docker volumes, and Docker containers.

Then, we applied all these concepts by building some of the ABC-MSA microservices with hands-on examples. We built the containers and showed how microservices communicate with each other.

In the next chapter, we will focus on building an AI microservice in the MSA system. We will discuss some of the most important AI/ML/DL algorithms that should be considered and implemented in an MSA system, and how these algorithms help with a system's overall stability, performance, and supportability.

10
Building an Intelligent MSA Enterprise System

In previous chapters, we gradually built the ABC-MSA to demonstrate some of an MSA system's features, techniques, and traffic patterns.

In this chapter, we will combine both MSA concepts and AI concepts to build an ABC-Intelligent-MSA system, which is an enhanced version of our ABC-MSA. The intelligent version of the ABC-MSA will use various AI algorithms to enhance the performance and general operations of the original ABC-MSA system.

ABC-Intelligent-MSA will be able to examine different traffic patterns and detect potential problems and then **self-rectify** or self-adjust to try to prevent the problem from taking place before it actually happens.

The ABC-Intelligent-MSA will be able to **self-learn** the traffic behavior, API calls, and response patterns, and try to **self-heal** if a traffic anomaly or problematic pattern is detected for whatever reason.

The following topics are covered in this chapter:

- The machine learning advantage
- Building your first AI microservice
- The intelligent MSA system in action
- Analyzing AI service operations

The machine learning advantage

There are many areas in our MSA where we can leverage AI to enhance the system's reliability and operability. We will focus our system on two main potential areas of enhancement. One is to enhance the system response in case of a microservice failure or performance degradation. The second area of enhancement is to add a proactive circuit breaker role.

As we discussed in *Chapter 3*, the circuit breaker pattern is used to prevent a system cascading failure when one of the system's microservices fails to respond to API consumer requests promptly. Should a microservice fail or perform poorly, our AI will try to take proactive action to fix the problem rather than waiting for the problem to be manually fixed for the system to return to normal operation.

In *Chapter 7*, we discussed the advantages of using **Machine Learning** (**ML**) and DL in MSA in detail. This chapter will focus on building two AI microservices to enhance our MSA system.

The first AI microservice is called a **Performance Baseline Watchdog** (**PBW**) service. The PBW is an ML microservice that creates a baseline for the expected performance of each microservice in the MSA system under a certain system load. Should the operational performance of the measured microservice fall under the performance baseline by the configurable value of x, the system should send a warning message to the **Operation Support System** (**OSS**) or the **Network Management System** (**NMS**) and should performance fall by y (which is also configurable), the system then should take predefined action(s) to try to self-rectify and self-heal the MSA system.

The second AI microservice we will build in this chapter is the **Performance Anomaly Detector** (**PAD**) service. The PAD is an ML microservice that takes a holistic view of the entire MSA system. The PAD learns the MSA performance patterns and tries to detect any anomalous behavior. It identifies "problematic patterns," tries to automatically detect a problem before it happens, and accordingly takes proactive action to fix the faulty area of the system.

Building your first AI microservice

Before we start building our two AI microservices, we need to think about our training and test data first – how we will collect our training data, build the model accordingly, test the model and measure its reliability, and enhance the algorithm's reliability if needed.

> **Important Note**
>
> The AI services we are building in our MSA system are only a proof of concept to demonstrate the value of implementing AI in MSA systems. Rather, businesses should consider an AI service or model that matches their unique needs, business process, and their deployed MSA system.

We will also need to simulate the use cases themselves. Simulate a system's microservice failure or performance degradation, simulate a cascading failure, and we should also be able to simulate some system's outlier patterns to see how the algorithm would detect and react to pattern anomalies.

To do all this, let's first understand how the PBW and PAD microservices fit with the overall system's operation and how they would normally interact with the different system's components.

The anatomy of AI enhancements

The main role of both the PBW and PAD is to enhance the stability and reliability of our MSA system. It is therefore imperative for both services to constantly watch individual microservices and the overall system performance and then take the necessary action when performance issues are detected.

The training data is first collected in a controlled environment for a specific training period, where normal, stable system operations and the average user load are simulated and applied. This can be achieved using some of the simulation tools we built, which will be discussed later in this section.

This training period creates an ideal first baseline that will be the main reference for the AI services to use during actual production time. The collected training data will then be used to build the algorithm. To achieve better and more accurate results, the training data and algorithm can be regularly tuned later when more information about real-time production traffic is collected.

The simulated load and system operations are tweaked through multiple simulation parameters. These parameters are tweaked regularly to mimic the actual acceptable operational performance. The algorithm tweaks would eventually stop (or become very minor) as the AI algorithms mature. The cycle of onboarding the AI services to the ABC-MSA system is demonstrated in *Figure 10.1*:

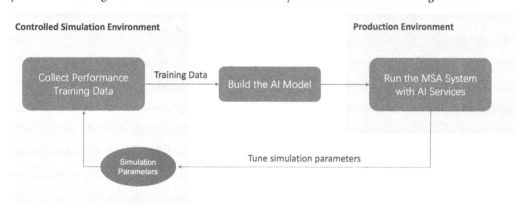

Figure 10.1: AI microservices implementation in ABC-Intelligent-MSA

More information on the simulation tools and parameters is coming up in the next section.

Once the AI services are operational, they will start collecting performance stats from each of the system's microservices through periodic API calls and then compare these performance stats with the expected performance or behavior.

Should an individual microservice or the overall system performance deviate from what the AI expects to see, a system action will be triggered to either warn the system administrators or self-heal whenever possible. *Figure 10.2* shows the high-level architecture of the PBW and PAD services:

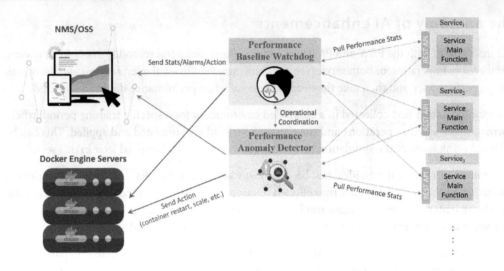

Figure 10.2: PBW and PAD services in ABC-Intelligent-MSA

The PBW's algorithm calculates the expected performance metrics based on the performance stats collected. Collected performance stats include API call response time stats, the failures or failure rate of individual microservices, the API response code, and the load applied on the microservice itself.

Pre-defined actions are triggered based on how far the microservice deviates from the calculated performance metric. Based on the configuration of the PBW, the higher the deviation, the more likely a proactive action is to be triggered to try to self-heal. In the case of a slight deviation, however, no healing action is supposed to be trigged; a system warning informing the system administrator is sufficient.

The following table shows some of the possible system issues that could be encountered during the operations of an ABC-Intelligent-MSA system, and the actions the PBW service would take to try to rectify the problem.

The list shown in the table is only a sample of potential issues and can, of course, grow as more use cases are considered:

Microservice Issue	Triggered Action
Slow responsiveness	Scale the microservice vertically/horizontally or restart the microservice
Intermittent timeouts	Scale vertically/horizontally or restart
API call HTTP response errors	Check Apache, Flask, the JVM, the Docker volume, SQL service, etc. Restart the service if needed Restart the microservice's container
Service is unresponsive (down)	Restart the microservice's container

Table 10.1 – Potential ABC-Intelligent-MSA operational issues and the PBW's self-healing actions

The healing mechanism can be applied to the MSA system using multiple AI services, not necessarily only using the PBW and PAD that we are implementing in our ABC-Intelligent-MSA. This is just an example.

The self-healing process

All of the PBW's healing actions listed in *Table 10.1* should not be taken in isolation from the PAD's operations, but rather should be carefully coordinated with the PAD's healing actions. A single issue in a microservice could (although not necessarily) trigger actions from both the PBW and PAD services at the same time and could consequently create an operational conflict.

In terms of the self-healing process, and to avoid conflict between the system's AI services when triggering self-healing actions, whenever an action is determined and before it is triggered, the AI service sends an API call to the other AI services first (either directly or through the API gateway), declaring a **Self-Healing Lock State** in the troubled microservice. Accordingly, all the other AI services in the MSA system will hold off any actions that may have been planned related to that troubled microservice.

During the self-healing lock state, the only AI service allowed to work on the troubled microservice is the **Healer AI Service**, which is the AI service that locked it.

Once the healer has fixed the problem and detects a normal operation in the affected microservice, the healer then sends another API call to the other AI services in the MSA system declaring that the lock state is over.

If the healer is unable to self-heal and gives up on resolving the issue, it sends an alarm to the NMS/OSS and marks that microservice as **unhealable** for a specific configurable period of time, known as the **Unhealable Wait Period** (by default, 15 minutes).

The unhealable wait period allows other AI services to try to heal that microservice and gives the healer a breather to pace out its operation across all other microservices in the MSA system.

To prevent healers from consuming system resources by slipping into indefinite healing attempts, healers will try to heal the troubled microservices for a specific number of healing attempts, configured through the **Maximum Healing Attempts** value (four attempts, by default), and will then completely give up trying. If the maximum healing attempts are exhausted, a manual system intervention will be needed to fix the troubled microservice.

System administrators can still configure indefinite healing attempts if needed, but this can consume system resources and may not be effective depending on the nature of the problem the MSA system or a specific microservice is experiencing.

If another AI service can fix the troubled microservice or the microservice is manually fixed, the original healer will automatically clear the unhealable flag of the microservice after the unhealable wait period is over.

If on the other hand, no other AI service can fix the problem and no manual intervention is taken to fix the microservice, the original healer – and any other healer that may have tried to fix the microservice – will try to heal the microservice again once the unhealable wait period expires if and only if the troubled microservice is not in a self-healing lock state.

The following visual chart summarizes the self-healing process and may help better explain the entire process.

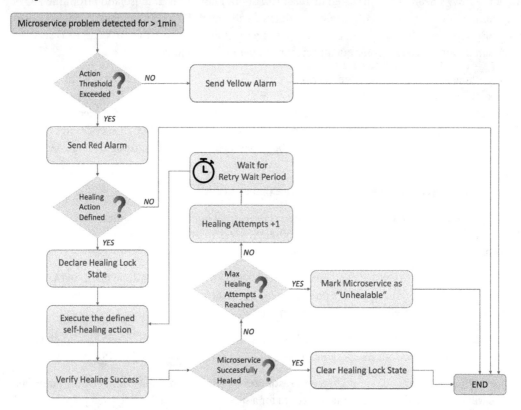

Figure 10.3: The self-healing process in MSA

It is important to also understand the main terminology used to explain the self-healing process. The following table shows a summary of the terminology of the main components of our self-healing process:

Term	Description
Healer	An AI service that attempts to heal a troubled microservice.
Healing Action	An action taken by the healer to try to fix an ongoing system operational issue.
Self-Healing Lock State	A microservice state in which an attempt is made by the healer to fix the microservice. In this state, only one healer (the one that initiated the lock state) is allowed to work on the troubled microservice. A microservice self-healing lock state is a state visible by the entire MSA system and not a healer-specific state.
Retry Wait Period	The time the healer for which has to wait when a healing action fails before it retries. The Retry Wait Period is 2 min by default.
Unhealable State	The state in which the troubled microservice is marked unfixable by a healer after a healer's failed attempt to fix the troubled microservice. A microservice unhealable state is a healer-specific state and only visible to the healer that gave up on fixing that troubled microservice. Other healers can still try to fix the troubled microservice.
Unhealable Wait Period	The time for which the healer has to wait before it starts to make another attempt to fix the troubled microservice. The Unhealable Wait Period is 15 min by default.
Maximum Healing Attempts	The maximum number of attempts the healer will try after each unhealable wait period, and before the healer totally gives up on the troubled microservice and no longer attempt to fix it. By default, PBW tries 4 healing attempts.

Table 10.2: ABC-Intelligent-MSA operational issues with the self-rectifying actions of the PBW

So far, we have explained the value of deploying AI services in the MSA system and shown some practical application examples to demonstrate the value of AI in MSA.

In order to build, run, and tweak AI services in MSA, we need to build certain tools to gather and log system statuses, operational dynamics, and operational statistics. In the following section, we will dive into what these tools are and how to use them.

Building the necessary tools

The purpose of creating project tools is to first be able to build the AI models, then simulate the entire ABC-Intelligent-MSA system, and then collect stats and analyze the system's operations.

Although there may be tools available online that would help us achieve our purpose, instead, we will build simple tools customized specifically for our use cases.

We created multiple tools to help us collect training and test data, simulate the system and microservices load, and measure the performance of the microservices. All the tools are available in the `tools` directory in our GitHub repository.

The tools also help us scrub some of the generated logs and data for analysis and potential future enhancements.

The following are the main tools we need in our ABC-Intelligent-MSA setup.

An API traffic simulator

The API traffic generator/simulator, `simulate_api_rqsts.py`, helps simulate the API request load for one or more of the system's microservices.

`simulate_api_rqsts` creates multi-threaded API requests across multiple target microservices. API HTTP requests are then sent to each microservice in parallel.

The API load is measured by requests per minute and API requests can either be uniformly or randomly paced.

The uniformly paced requests are paced out so that the time between each API call is always the same, so if we are configuring a uniformly paced load of 600 API requests/min, `simulate_api_rqsts` will send 1 API call every T = 100 ms.

In the randomly-paced case, each API call is sent after a random period, T_R, from the time where the previous call was sent, but so that T_R can never be larger or smaller than 95% of T. So if we are configuring a randomly-paced load of 600 API requests/min, T_R, in that case, will be equal to a value greater than 5 ms and smaller than 195 ms.

`simulate_api_rqsts` will send 1 API call every:

$(1-95\%)T <= T_R <= (1+95\%)T$ (i.e., for 600 requests/min: *5 ms <= TR <= 195 ms*)

The sum of all T_Rs, however, will still be approximately equal to the configured requests/min. In our example here, the load is 600 API requests/min.

Uniformly paced requests are better when you are manually analyzing how a particular microservice responds to the API load, while randomly paced requests are a better representation of a real-time production API request load.

The microservices performance monitor

The microservices performance monitor, `ms_perfmon.py`, is another multithreading tool and is initially used for collecting and building the AI training data during the simulation period of ideal conditions.

`ms_perfmon` sends parallel API calls to each microservice in the system and then logs the API call hyperlink, the date and time at which it was sent, the receiving microservice response time, and the HTTP response code. The following is an example log entry of the collected data in a comma-separated format:

```
http://payment_ms:8080,2022-12-28 15:48:57.271370,
0.010991334915161133,200
```

Each microservice stat is collected in its own **Comma-Separated Values** (**CSV**) log file named after the API link itself (after cleaning up special characters). All the stat files are collected under the `perfmon_stats` directory in the `ms_perfmon` working path.

In real-time operation, both the PBW and PAD perform a similar job to `ms_perfmon`. They collect their own stats and measure the target microservice's real-time performance against the baseline and the expected normal behavior.

Should we extend the MSA system's AI capabilities by including more AI services for different purposes and use cases, which will likely require each AI service to conduct its own performance statistics collection?

Depending on the collection frequency and the type of data collected, as the number of collectors increases, scalability could become an issue. The `ms_perfmon` function, in that case, can be extended to become the main AI collector for all AI or non-AI services in the MSA system. This setup can help offload the system's microservices and allow the MSA system to scale better.

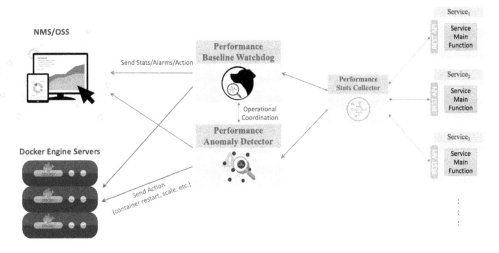

Figure 10.4: A collect-once performance stats setup in ABC-Intelligent-MSA

Figure 10.4 shows how ms_perfmon can handle stats collection on behalf of all other services in the MSA system and then act as a proxy and respond to API calls requesting whatever stats are needed for each particular AI (or non-AI) service.

The response delay simulator

To simulate a delayed response or a troubled microservice, and solely for simulation and testing purposes, we added a feature in key microservices to simulate a delayed API call response.

The delay response feature, when enabled in the microservice, has two configurable values – the minimum delay and the maximum delay. When a microservice receives an API call, it will automatically assign a random delay value between the configured minimum delay and maximum delay, and then wait for that time before it responds to the consumer's API calls.

The feature is very helpful for simulating a cascading system failure. As will be shown later in this chapter, the response delay feature can also help demonstrate the value of using AI services to enhance the operations of the MSA system compared to using the short circuit traffic pattern previously explained in *Chapter 3*.

The response delay is enabled whenever the max delay is configured with a value greater than zero. When the value of max delay is higher than zero, a delay value is assigned to the microservice API's call response, as shown in the following code snippet:

```
#Simulate a delay if received an API to do so
if delay_max_sec > 0:
   delay_seconds = round(delay_min_sec + random.random()*(delay_
max_sec-delay_min_sec), 3)
   #print("Adding a delay %s ..." %delay_seconds)
   time.sleep(delay_seconds)
```

The max and min delay values can be configured using an API call. The following is an example of using curl to send an API call to configure the maximum and minimum delay response in milliseconds:

```
curl http://inventory_ms:8080/api/
simulatedelay?min=1500&max=3500
```

Again, this feature is only for demo and test purposes. A more secure way of simulating a delay is using secured configuration files or local parameters instead.

The API response error simulator

Similar to the response delay simulator, this feature is for demo purposes only. The API error simulator feature uses one configurable value – the average HTTP error per hour. When the feature is enabled in the microservice, the microservice will pick a randomly applicable server 500 error and respond to API requests with randomly paced responses that match the configured error rate.

The error rate can be configured using an API call. The following is an example of using `curl` to send an API call to configure an API error response rate of 5 HTTP errors per hour:

```
curl http://inventory_ms:8080/api/response_err?rate=5
```

Now, we know the testing and simulation tools available for us to use for training, testing, and simulating production for our MSA system.

In the next section, we will discuss our ABC-Intelligent-MSA operations – how to initialize the system, how to build and use training and testing data, and how to simulate the system's production traffic.

The intelligent MSA system in action

In the previous sections of this chapter, we discussed how the different system components interact with each other and what tools we use to build the AI algorithms, test the system, and monitor the operations of different components.

In this section, we will put our ABC-Intelligent-MSA to the test. We will run all system microservices and tools, and see how the different system components actually interact with each other, what results we see, and how we can tweak the system to maintain smooth end-to-end operations.

The ABC-Intelligent-MSA will first run under an ideal simulation environment (no error simulation and no delays) to collect the training data necessary to build the AI models. Once enough data has been collected, we will then train the models and prepare the system for actual production traffic.

The system initialization steps, therefore, are as follows:

1. Start the system with no AI services to collect the necessary training data under an ideal operational situation and create an operational baseline.

2. Sanitize the collected data if needed and remove outliers.

3. Train the AI algorithms using the training data collected.

4. Re-initialize the system with all of its AI services.

5. Start production operations. In our example here, we will simulate actual production operations by injecting errors, data delay responses, service failures, and so on.

Initializing the ABC-Intelligent-MSA system

We start by initializing our MSA system using the system's Docker compose file, abc_ msa.yaml, and using the docker-compose command as follows,

```
$ docker-compose -f abc_msa.yaml up &
```

As discussed previously in *Chapter 9*, the preceding docker-compose command is much more convenient than using multiple docker run commands. docker-compose will read the system's run parameters and configuration from the abc _msa.yaml file, and initialize all the system components accordingly.

In our example, this will start the analysis and monitoring tools, along with all the regular microservices in the system. Since we are still collecting training data, no AI services need to be initialized yet.

As shown in *Figure 10.2* and *Figure 10.4*, when we start the AI services (the PBW and PAD), they will need to be able to remotely control (start, stop, and restart) the system's Docker containers. The PBW and PAD are designed to control the Docker containers using API calls. Therefore, we need to enable Docker Engine first to respond to API calls and for the PBW and PAD to be able to successfully communicate with Docker Engine.

The following are the steps needed to enable Docker's API remote management:

1. On your Ubuntu system, use vi, vim, or any other similar tool to edit the /lib/systemd/system/docker.service file.

2. Look for the ExecStart entry and make the necessary modifications for it to be like the following:

    ```
    ExecStart=/usr/bin/dockerd -H=fd:// -H=tcp://0.0.0.0:2375
    ```

3. This will enable Docker Engine to listen to API calls. Make sure you save the file after the modifications.

4. Reload Docker Engine using the following command:

    ```
    systemctl daemon-reload
    ```

5. To ensure Docker Engine is working properly and responding to API calls, use the following command:

    ```
    curl http://localhost:8080/version
    ```

Now, the system is running and collecting training data. The longer you run the system, the more training data will be collected, and the more accurate your AI models will be. In our example, we will leave the system running for approximately 48 hours.

In the next subsection, we will go over how to run the tools, build training data, collect some of the system performance logs, simulate real-time system operations, and analyze the collected performance data.

Building and using the training data

The `ms_perfmon` tool will create a separate stat file for each microservice in the `<ms_perfmon's working path>/perfmon_stats` directory. It is important that we leave the tool running and monitor the system's performance stats under minimal load conditions.

We recommend at least 48 hours of training data collection. Ideally, however, data should be collected with seasonality load whenever applicable. In some environments, for example, the system load may increase on weekends over weekdays, during the shopping season, and so on. These situations should be considered in the training data to be able to build a more accurate AI model.

Performance data is pulled every 10 seconds, and accordingly, with 48h of active monitoring, `ms_perfmon` produces 17,280 entries for each microservice.

Regardless of the length of the system's training period, whenever enough performance data has been collected, the `training_data_cleanup.py` tool should be run to detect any outliers and sanitize the performance data before using it in our AI services.

The `training_data_cleanup` tool scrubs all the performance data files in the `<ms_perfmon's working path>/perfmon_stats` directory, and automatically creates a `scrubbed_stats` directory with all the scrubbed data for each microservice. These scrubbed files are the files that we will later use for training the AI services.

We are now ready to write our Python code for training the PBW:

1. We will use the `numpy` library for array and scientific data processing, `pandas` for reading our CSV training data files and testing data, and `sklearn` to build our AI model:

   ```
   import numpy as np
   import pandas as pd
   from sklearn.model_selection import train_test_split
   from sklearn.linear_model import LinearRegression
   ```

2. After importing the required libraries, we now need to copy all performance data into a DataFrame object. The following is a code example of this:

   ```
   payment_ms_stats_df = pd.read_csv('scrubbed_stats/
   payment_ms_stats.csv')
   ```

The PBW's AI model includes the microservice response time, the calculated request failure rate, and the calculated microservice load. The model should calculate the expected response time based on all the preceding parameters.

3. In our Python code, we need to point to the data column that needs to be predicted. In our example, that would be the response time. The following is a code snippet for the Payment microservice:

```
payment_ms_rt = np.array(payment_ms_stats_df['response_
time'])
```

4. We need now to build our model, but before doing so, we need to load the rest of the performance data column into an array for training and testing processing. We do that by removing the "response time" column (an axis of 1) from the created DataFrame and then loading that DataFrame into an array to be used in our sklearn object, as follows:

```
model_data = payment_ms_stats_df.drop('response_time',
axis = 1)
model_data = np.array(model_data)
```

5. The model data need to be split into training data and test data. We split the model data into 80% training and 20% test data as follows:

```
model_data_train, model_data_test, payment_ms_rt_train,
payment_ms_rt_test = train_test_split(model_data,
payment_ms_rt, test_size = 0.20, shuffle=True)
```

6. Now, we build the model from the training data:

```
lr_model = LinearRegression()
lr_model.fit(model_data_train, payment_ms_rt_train);
```

7. Save the data to a CSV file for future use:

```
trend_payment_ms_rt_predictions = lr_model.
predict(payment_ms_rt)
df = payment_ms_rt_df.assign(predicted_payment_ms_rt =
trend_payment_ms_rt_predictions)
df.to_csv("predicted_payment_ms_rt_trend.csv", mode =
'w', index=False)
```

Now, we have the training data and the trained model. It is time to use the model for production traffic.

In the next subsection, we will simulate production operations and describe how that can be applied to our trained MSA system, the ABC-Intelligent-MSA.

Simulating the ABC-Intelligent-MSA's operation

We need to reinitialize the system now with the trained model and production traffic. Since no actual production traffic is applied in our example, we need to simulate the production operation with its potential operational challenges, including high traffic loads, service failures, and potential network hiccups.

We start by reinitializing the ABC-Intelligent-MSA system using `docker-compose`, as described earlier, but using the `abc_intelligent_msa.yaml` file:

```
$ docker-compose -f abc_intelligent_msa.yaml up &
```

The main difference between `abc_intelligent_msa.yaml` and `abc _msa.yaml` is that the first file includes the initialization of the AI services.

Once the system is running, the AI tools will start monitoring and collecting the microservice's performance and trigger healing actions whenever a system problem is detected and metrics exceed the configured performance thresholds.

The production traffic is ready to be simulated now using the `simulate_api_rqsts` API traffic simulator and the response delay simulator function discussed earlier.

Using the API response error simulator, occasional HTTP errors can also be simulated if needed. A more sophisticated simulation would involve injecting HTTP 500 error codes as well, but we will stick to response time performance delays for simplicity.

The `ms_perfmon` tool will still be running to collect data for our offline analysis whenever needed.

We now need to simulate specific production use cases and see how the AI tools will respond and self-heal the entire system. In the next section, we will discuss the operations of the PBW and PAD and look into how both AI services interact with system performance readings and errors.

Analyzing AI service operations

In the preceding sections, we started by building our first AI service and covered how to use AI to enhance the MSA system's operations and resilience, the self-healing process, and the tools we built to generate training data and simulate the ABC-Intelligent-MSA system's operation.

In this section, we will examine the system logs and check how the PBW and PAD interact with the system and actually enhance its operations. We will then simulate a cascading system failure and examine how the self-healing process is triggered and handled to bring the MSA system back to normal operation.

The PBW in action

During the training period, the PBW was able to build an AI model and calculate the expected response time of each microservice in the ABC-Intelligent-MSA system. As you can see from the following log sample, under a normal system load, the average response time of the Inventory microservice is about 20 ms:

```
http://inventory_ms:8080,2022-11-23 15:48:25.094675,
0.01450204849243164,200
http://inventory_ms:8080,2022-11-23 15:48:35.816913,
0.0241086483001709,200
http://inventory_ms:8080,2022-11-23 15:48:46.543205,
0.02363872528076172,200
http://inventory_ms:8080,2022-11-23 15:48:57.271370,
0.010991334915161133,200
http://inventory_ms:8080,2022-11-23 15:49:07.983282,
0.021454334259033203,200
http://inventory_ms:8080,2022-11-23 15:49:18.645113,
0.012285232543945312,200
http://inventory_ms:8080,2022-11-23 15:49:29.310656,
0.0245664119720459,200
http://inventory_ms:8080,2022-11-23 15:49:40.010556,
0.013091325759887695,200
http://inventory_ms:8080,2022-11-23 15:49:50.744695,
0.021291017532348633,200
http://inventory_ms:8080,2022-11-23 15:50:01.715555,
0.024635791778564453,200
```

We configured the warning threshold for the PBW as 250 ms, and the action threshold as 750 ms. We will now start introducing an API call load to the Inventory microservice using `simulate_api_rqsts` and delays using the response delay simulator feature. Then, we will see how the PBW reacts from the PBW action logs.

The following are the PBW's performance readings for about 1.5 minutes. As you can see from the readings, the response time is consistently above the 250 ms alarm threshold, but (with the exception of one reading) still below the 750 ms action threshold:

```
http://inventory_ms:8080,2022-11-23 18:24:00.518005,
0.6386377334594727,200
http://inventory_ms:8080,2022-11-23 18:24:11.469172,
0.7164063453674316,200
http://inventory_ms:8080,2022-11-23 18:24:22.203452,
```

```
0.7233438491821289,200
http://inventory_ms:8080,2022-11-23 18:24:32.942619,
0.7101089954376221,200
http://inventory_ms:8080,2022-11-23 18:24:43.668907,
0.6982685089111328,200
http://inventory_ms:8080,2022-11-23 18:24:54.777383,
0.8207950115203857,200
http://inventory_ms:8080,2022-11-23 18:25:05.410204,
0.6812236309051514,200
http://inventory_ms:8080,2022-11-23 18:25:16.101344,
0.6544813632965088,200
http://inventory_ms:8080,2022-11-23 18:25:27.072040,
0.7446155548095703,200
http://inventory_ms:8080,2022-11-23 18:25:37.828189,
0.6969136238098145,200
```

The readings will have to be consistently above the 750 ms action threshold for the PBW to trigger a healing action. One reading above 750 ms is not enough for an action to be triggered. However, since the readings are constantly above the 250 ms alarm threshold, the PBW is expected to trigger an alarm to the NMS/OSS system.

We need to verify the PBW's behavior from the NMS/OSS system or the PBW's action log. The following is a snippet of the PBW's action log during the same period from the previous example:

```
2022-11-23 18:24:00.518005: Alarming high response time
(0.6386377334594727) detected in inventory_ms. No alarm
triggered yet.
2022-11-23 18:24:11.469172: Alarming high response time
(0.7164063453674316) detected in inventory_ms. No alarm
triggered yet.
2022-11-23 18:24:22.203452: Alarming high response time
(0.7233438491821289) detected in inventory_ms. No alarm
triggered yet.
2022-11-23 18:24:32.942619: Alarming high response time
(0.7101089954376221) detected in inventory_ms. No alarm
triggered yet.
2022-11-23 18:24:43.668907: Alarming high response time
(0.6982685089111328) detected in inventory_ms. No alarm
triggered yet.
2022-11-23 18:24:54.777383: Actionable high response time
(0.8207950115203857) detected in inventory_ms. No action
triggered yet.
```

```
2022-11-23 18:25:05.410204: Alarming high response time
(0.6812236309051514) detected in inventory_ms. No alarm
triggered yet.
2022-11-23 18:25:16.101344: Alarming high response time
(0.6544813632965088) detected in inventory_ms. No alarm
triggered yet.
2022-11-23 18:25:27.072040: Alarming high response time
(0.7446155548095703) detected in inventory_ms. No alarm
triggered yet.
2022-11-23 18:25:37.828189: Alarming high response time
(0.6969136238098145) detected in inventory_ms. Yellow alarm
triggered and sent to NMS/OSS system.
2022-11-23 18:25:48.637317: Alarming high response time
(0.6777710914611816) detected in inventory_ms. Yellow alarm
triggered and sent to NMS/OSS system.
2022-11-23 18:25:59.327946: Alarming high response time
(0.6758050918579102) detected in inventory_ms. Yellow alarm
triggered and sent to NMS/OSS system.
2022-11-23 18:26:10.014319: Alarming high response time
(0.6641242504119873) detected in inventory_ms. Yellow alarm
triggered and sent to NMS/OSS system.
```

As you can see from the preceding snippet's last 4 log entries, after a consistent delay of more than 250 ms, an alarm was triggered and sent to the NMS/OSS system. We need to increase the inventory microservice's load and response time to see how the PBW will react.

The following is another snippet of the PBW's performance log. Only the last 4 log entries in a series of 10 consistent response delay readings are above 750 ms:

```
http://inventory_ms:8080,2022-11-23 18:29:31.852330,
1.326528787612915,200
http://inventory_ms:8080,2022-11-23 18:29:43.196200,
1.4279899597167969,200
http://inventory_ms:8080,2022-11-23 18:30:05.310226,
1.0108487606048584,200
http://inventory_ms:8080,2022-11-23 18:30:16.334608,
1.1380960941314697,200
```

Normally, we would have configured all healing actions shown in *Table 10.1*. In our demo system, however, we have configured only one healing action to demo the system self-healing operations in general. We only configured a microservice container to restart if a problem is experienced in the microservice. The response delay simulator feature is therefore a more relevant simulation tool than the other tools we have mentioned earlier.

In case of slow performance due to high API call requests volume, the most appropriate healing action would be to try to scale the microservice first and allocate more resources to respond to the high volume of API requests.

We assume in our simulation that the problem in the Inventory microservice is not necessarily due to the API request load, but rather some unforeseen problem causing the Inventory service to become unstable and unable to handle API calls promptly, so restarting the Inventory microservice could therefore fix the problem.

Now, here is a look at the PBW's action log during the same period. Please note that prior to the actionably high response time, an alarmingly high response time below 750 ms was previously detected. The response time was higher than 250 ms and below 750 ms:

```
2022-11-23 18:29:31.852330: Actionable high response time
(1.326528787612915) detected in inventory_ms. No action
triggered yet. Yellow alarm triggered and sent to NMS/OSS
system.
2022-11-23 18:29:43.196200: Actionable high response time
(1.4279899597167969) detected in inventory_ms. No action
triggered yet. Yellow alarm triggered and sent to NMS/OSS
system.
2022-11-23 18:30:05.310226: Actionable high response time
(1.0108487606048584) detected in inventory_ms. No action
triggered yet. Yellow alarm triggered and sent to NMS/OSS
system.
2022-11-23 18:30:16.334608: Actionable high response time
(1.1380960941314697) detected in inventory_ms. Red Alarm
triggered and sent to NMS/OSS system.
2022-11-23 18:30:16.334608: Self-healing lock state declared
for inventory_ms container.
2022-11-23 18:30:16.334608: Self-healing action triggered.
Restarting inventory_ms container (inventory_management_
container).
2022-11-23 18:30:21.359377: Verifying inventory_ms
operations...
2022-11-23 18:30:22.945823: inventory_ms was successfully
restarted
2022-11-23 18:30:23.089051: Self-healing lock state cleared for
inventory_ms container.
```

As you see from the last 4 entries in the action log, the PBW detected a consistent response time (above 750 ms) and accordingly sent a red alarm to the NMS/OSS system, indicating a critical delay in the Inventory service and the need for a self-healing action to be taken. The PBW then locked the Inventory microservice to avoid clashing with healing actions from other AI services. The PBW then restarted the Inventory microservice by sending a restart API call to Docker Engine, verified that the Inventory microservice was back online, and finally unlocked the Inventory microservice.

To restart a Docker container through API, you will need to send a POST request as follows:

```
/containers/<container id or name>/restart
```

You can also specify the number of seconds to wait before restarting the container using a t parameter. The following is a container restart POST example to restart the Inventory service container after a 10-second wait time:

```
/v1.24/containers/inventory_management_container/restart?t=10
```

For more information on how to control Docker Engine using API calls, check the Docker Engine API documentation at https://docs.docker.com/engine/api/version-history/.

However, was the PBW able to fix the Inventory microservice problem?

Let's go back now to the PBW's performance log and see how this self-healing action impacted the Inventory service performance. The following are the log entries just before the healing action was triggered:

```
http://inventory_ms:8080,2022-11-23 18:30:16.334608,
0.1380960941314697,200
http://inventory_ms:8080,2022-11-23 18:30:27.629649,
0.1693825721740723,200
http://inventory_ms:8080,2022-11-23 18:30:38.486793,
0.1700718116760254,200
```

Sure enough, the response time dropped from above 1 s to a maximum of 170 ms. Not as low as it was before the problem appeared, but the Inventory microservice for sure has some breathing room now. The performance issues may very well return if the underlying problem is not attended to and properly fixed.

In a more advanced AI model, we can train and configure the system to take more sophisticated actions to fully resolve the problem whenever needed, but in this book, we are limited to a specific scope to be able to demonstrate the idea in principle and pave the way for you to develop your own AI models and algorithms for your specific use cases.

We have demonstrated in this section how the PBW works and how an action is triggered when a microservice performance issue is detected. In the following section, we will go over the PAD AI service and how the PAD takes a rather more holistic view of the entire system.

The PAD in action

The best way to demonstrate the operations of the PAD is to simulate a cascading failure and see how the PAD can bring the MSA system back to normal operation.

To simulate a cascading failure and ensure that the PAD responds to the failure and tries to auto-heal, we will first need to disable the PBW AI service. This will prevent the PBW from triggering a healing action and prevent it from trying to resolve the problem before the PAD's healing action(s) kick in.

Let's quickly revisit what we have previously discussed in *Chapter 3*, an example of how a cascading failure happens.

As shown in *Figure 10.5*, under heavy API traffic, a failure to the Inventory microservice could cause the **Payment** microservice to pile up too many API calls in the queue, waiting for a response from the Inventory service. Eventually, these API calls will consume and exhaust the available resources in the **Payment** microservice, causing it to fail. A failure in the Payment microservice will produce a similar situation in the **Order** microservice, and eventually, produce a failure for the **Order** microservice as well:

Figure 10.5: The Payment microservice is down

For the PAD to respond with healing actions, each of the PAD's detected anomaly types has to have healing actions defined for it.

To successfully simulate the cascading failure, we only defined an action for a cascading failure situation. Otherwise, the PAD would automatically detect the failure in the Inventory service and self-heal it by restarting the Inventory microservice container, preventing a cascading failure from happening to begin with.

We will start by simulating a high volume of orders for the Order microservice and see how the system is going to respond to this situation in general, and specifically how the PAD will react under the situation.

To simulate a high volume of order requests, use the following `simulate_api_rqsts` command to target the Order microservice with a fixed uniformly paced order requests of 100,000 per minute:

```
simulate_api_rqsts 100000 http://order_ms:8080/place_order
```

We will now shut down the Inventory microservice and examine the PAD action logs. The following is a snippet of the log about a minute after the PAD started to detect a failure in the Inventory microservice.

Please note that we introduced sudden high-volume traffic into the system. This sudden traffic increase by itself is a traffic pattern anomaly that was picked up by the PAD, but the PAD did not respond to that specific anomaly because no healing action is specifically defined for that anomaly:

```
2022-11-24 11:39:13.602130: Traffic pattern anomaly detected,
(inventory_ms) is likely down. No action is defined. No action
triggered yet. Yellow alarm triggered and sent to NMS/OSS
system.
2022-11-24 11:39:23.469204: Traffic pattern anomaly detected,
(payment_ms) slow API response detected. No action is defined.
No action triggered yet.
   :

   :

2022-11-24 11:40:26.836405: Traffic pattern anomaly detected,
(payment_ms) slow API response detected. No action is defined.
No action triggered yet. Yellow alarm triggered and sent to
NMS/OSS system.
```

In the preceding snippet of the PAD log, the PAD automatically recognized the Inventory service failure since no response traffic was detected from the service. However, no action was taken by the PAD since no healing action was defined for that particular anomaly. Since the anomaly was consistent for more than 1 minute, the PAD sent an alarm to the NMS/OSS system to notify the system admins of the problem.

Because of the Inventory microservice failure, the Payment microservice started to run out of resources, and the PAD picked up an unusually slow traffic flow from the Payment microservice given the API call request load applied. Accordingly, and as seen in the log, a little over 1 minute later, the PAD started to generate alarms to NMS/OSS.

As shown in the following PAD log, a few minutes after the Payment microservice anomaly, the Order microservice started acting up, and accordingly, the PAD was able to correlate all these anomalies and detect a potential cascading failure:

```
2022-11-24 11:47:12.450897: Traffic pattern anomaly detected,
(order_ms) slow API response detected. No action is defined. No
action triggered yet. Yellow alarm triggered and sent to NMS/
OSS system.
2022-11-24 11:47:12.450897: Traffic pattern anomaly detected,
potential cascading failure detected. No action triggered yet.
Yellow alarm triggered and sent to NMS/OSS system.
```

Please note that the only microservice failure we have so far is the one we manually shut down, the Inventory microservice. Both the Payment and Order microservices are still up and running but, as it seems from the log, may be suffering from resource exhaustion.

The system is still running so far, and should the Inventory service return back online, the system will automatically recover. The user experience during the heavy load would only be slow performance during the ordering process, but no orders have been denied or failed yet.

By examining all these previously mentioned PAD action logs, and as the situation stands so far, we are still okay. However, if no action is taken to resolve the Inventory microservice problem, the system will eventually fail and user orders will start to be denied.

The short circuit traffic pattern discussed in *Chapter 3* helps prevent a cascading failure from taking place, but it still cannot resolve the underlying problem. User orders in a traditional short circuit pattern implantation will still be rejected until manual intervention fixes the Inventory microservice.

That's where the PAD comes in. Check the following PAD action log!

```
2022-11-24 11:48:13.638447: Traffic pattern anomaly detected,
potential cascading failure detected. (inventory_ms)
microservice is likely the root-cause. Red Alarm triggered and
sent to NMS/OSS system.
2022-11-24 11:48:13.638447: Self-healing lock state declared
for inventory_ms container.
2022-11-24 11:48:13.638447: Self-healing action triggered.
Restarting inventory_ms container (inventory_management_
container).
2022-11-24 11:48:18.663912: Verifying inventory_ms
operations...
2022-11-24 11:48:20.325807: inventory_ms was successfully
restarted
2022-11-24 11:48:20.474590: Self-healing lock state cleared for
inventory_ms container.
```

The PAD was able to detect the cascading failure before it actually happened, and was able to identify the root cause of the problem. The PAD sent a red alarm to the NMS/OSS system, declared a self-healing lock state on the Inventory service to try to fix the problem's root cause, was able to successfully restart the Inventory microservice container, and then cleared the self-healing lock on the Inventory service.

Let's now check the microservices performance logs and ensure that the problem is fixed and that the ABC-Intelligent-MSA system and all of its microservices are running normally.

Here's the Inventory microservice's performance log:

```
http://inventory_ms:8080,2022-11-24 11:51:33.132089,
0.033451717535487,200
http://inventory_ms:8080,2022-11-24 11:51:43.894705,
0.035784934718275,200
http://inventory_ms:8080,2022-11-24 11:51:54.809743,
0.027584526453594,200
http://inventory_ms:8080,2022-11-24 11:52:06.155834,
0.028615804809435,200
```

Here's the Payment microservice's performance log:

```
http://payment_ms:8080,2022-11-24 11:54:41.109835,
0.051435877463506,200
http://payment_ms:8080,2022-11-24 11:54:51.924508,
0.102346014326819,200
http://payment_ms:8080,2022-11-24 11:55:03.372841,
0.070163827689135,200
http://payment_ms:8080,2022-11-24 11:55:14.076832,
0.157682760576845,200
```

Here's the Order microservice's performance log:

```
http://order_ms:8080,2022-11-24 11:58:37.135827,
0.209097164508914,200
http://order_ms:8080,2022-11-24 11:58:47.584731,
0.193851625041193,200
http://order_ms:8080,2022-11-24 11:58:58.243759,
0.150628069240741,200
http://order_ms:8080,2022-11-24 11:59:08.961412,
0.138192362340785,200
```

As shown for the preceding Inventory, Payment, and Order microservices, all of those microservices are back online with normal performance readings. The system is now back to normal operation and should be able to handle the production load with no issues.

Summary

This chapter walked us through how we can build AI models to build an intelligent MSA system step by step. We accordingly built two main AI services – the PBW and the PAD – and leveraged these AI services to enhance our MSA demo system, ABC-MSA, to build an intelligent MSA system that we named ABC-Intelligent-MSA.

We explained the self-healing process design and dynamics in detail, as well as the tools we built to develop AI training data, how to simulate production operations, and how to measure the demo system's performance. We then put the ABC-Intelligent-MSA to test, simulated a couple of use cases to demonstrate AI functions within the MSA system, and carefully examined the logs of our demo AI services to showcase the value of using AI in MSA.

Everything explained in this chapter is just an example of using AI in an MSA system. Enterprises should consider using AI services that are specifically appropriate for their own MSA system and use cases. These AI tools may very well be available through third parties or built in-house whenever needed.

In the next chapter, we will discuss the transformation process from a traditional MSA system to an intelligent MSA system – the things to consider in greenfield and brownfield implementations, and how to avoid integration challenges to make the corporate transformation as smooth as possible.

11

Managing the New System's Deployment – Greenfield versus Brownfield

In the previous chapters, we discussed building an MSA system and integrating AI algorithms to form an Intelligent MSA. We covered concepts, techniques, and methodologies while accompanying them with examples.

In this chapter, we will discuss the different **greenfield** and **brownfield** deployment considerations, and ways to smoothly deploy the new intelligent MSA system with minimal operational disruptions, to be able to maintain overall system stability and business continuity.

We will also examine how to overcome general deployment challenges, particularly in brownfield deployments where existing systems are in production, and implement a successful and effective migration plan for the new Intelligent MSA system.

The following topics will be covered in this chapter:

- Deployment strategies
- Greenfield versus brownfield deployment
- Overcoming deployment challenges

Deployment strategies

Organizations utilize various techniques to minimize downtime and ensure a seamless and successful deployment when deploying a new system. Some of the most commonly used deployment strategies organizations follow are Recreate, Ramped, Blue/Green, Canary, A/B Testing, and Shadow deployments:

- The **Recreate deployment** is a simple, straightforward approach that involves replacing the entire infrastructure at once, similar to the Big Bang migration we discussed in *Chapter 3*. This approach is best suited for small and simple systems; however, it also means that the system is completely offline during the deployment process, which can lead to significant downtime.

- The **Ramped deployment** is similar to the Trickle migration we discussed in *Chapter 3*. The Ramped deployment allows the existing system to remain online during the deployment process. The new system is brought online gradually, and traffic is gradually routed to it, allowing both systems to stay available to users throughout the deployment process. Although this can be effective in small businesses and simple systems, this approach is ideal for larger, more complex systems, where minimizing downtime is a priority.

- **Blue/Green deployment** is a technique that involves maintaining two identical production environments, referred to as "blue" and "green," and routing traffic to one or the other. This allows for a seamless switchover in case any operational issues are experienced in the newly deployed version. This method is best suited for mission-critical systems since it ensures that there is always a working system available to users at any given time.

- **Canary deployment** is a technique that involves deploying the new system alongside the existing one and routing a small percentage of traffic to the new system. This allows for testing the new system with actual production traffic before rolling it out completely. If problems arise in the new system, the rollout can be reevaluated based on the type of issues encountered; then, the previous system can be reinstated while the problems are being resolved. This approach is often used to deploy changes to critical systems that require high levels of availability.

- The **A/B testing deployment** is another approach that involves simultaneously running both the old and the new system but testing them with different subsets of users to determine which performs better. This method works best for testing new system features or services.

- In the **Shadow deployment**, the new system is deployed to run alongside the existing system. The live production traffic of the old system is then redirected to the new system to test a newly released feature, the system stability under load, or test the new system altogether. This approach is best used in large systems deployed in large organizations.

The following is a comparison summary of all the preceding deployment strategies:

	Continuous Uptime	Production Traffic Testing	Cost	Complexity
Recreate	No	No	Low	Low
Ramped	Yes	No	Low	Low
Blue/Green	Yes	No	High	Medium
Canary	Yes	Yes	Low	Medium
A/B Testing	Yes	Yes	Low	High
Shadow	Yes	Yes	High	High

Table 11.1: Comparison of the different deployment strategies

Each of these strategies has its pros and cons, and none of them would be best suited for every case. Organizations must choose the appropriate strategy based on their specific needs and the nature of the changes being deployed.

The complexity and design of the existing system that's being upgraded or replaced, as well as its age and operation and the technology stack being used, play significant roles in determining the deployment strategy. In the next section, we will discuss the greenfield and brownfield deployments and their impact on determining the specifics of the deployment approach and plan.

Greenfield versus brownfield deployment

With our intelligent MSA system ready for deployment, we now need to think in detail about the infrastructure we have or need to acquire to deploy the system in production.

Some of the main questions we need to address are as follows:

- What are the infrastructure details needed to run the intelligent MSA system?
- Do we have the hardware and software resources needed to deploy and run the system efficiently?
- Can we leverage our existing infrastructure and applications to deploy the new system?
- What is the delta between the infrastructure needed and the infrastructure we have in place, and how can we fill that gap?

The organization's current infrastructure setup and existing systems (if any) play a crucial role in answering all of the preceding questions – that is, whether the new system is being deployed in a greenfield or brownfield environment.

Greenfield deployment refers to building and deploying a new system or infrastructure from scratch with no previous system or infrastructure in place. Thus, we must build the new system without major constraints, dependencies, integration work, or compatibility issues to consider before building and running the new system.

On the other hand, a brownfield deployment refers to the process of deploying a new system or infrastructure on a site that is already running with an existing system in place. The site may have existing infrastructure such as servers, applications, network components, and more that may be reused for deploying the new system.

In short, greenfield deployment is a new start from scratch, whereas brownfield deployment involves building on top of existing systems or infrastructure and possibly dealing with some integration issues, compatibility concerns, and resource constraints.

Whether it is a greenfield or brownfield deployment is often determined by the organization's situation and how the organization's business process is being conducted. Nevertheless, it is still important to understand the pros and cons of each deployment type to plan accurately.

> **Important Note**
>
> If cost saving is a major focus in the organization, we should leverage as many existing components as possible from the existing brownfield infrastructure. However, we need to reuse existing components in a way that cannot negatively impact the deployed system's efficiency, reliability, or functionality.

The following are some factors to consider when evaluating both deployment types.

Flexibility

Because in greenfield deployment we are starting from scratch, this gives us the liberty to design, implement, and optimize the new system for the specific needs of the organization without restricting ourselves to any dependent components or existing production systems.

In brownfield, on the other hand, we must always think of the already running systems and their dependencies before designing or deploying any part of the new system. This in itself can limit the deployment, customization, or optimization of the new system.

Scalability

Greenfield implementations offer higher scalability compared to brownfield because, in greenfield implementations, we deploy new infrastructure without any existing constraints that could limit the design or customization of the new system.

This lack of constraints gives architects and system designers the choice to design the system so that it scales without thinking of underlying technology or existing equipment that may hinder the system's capabilities or scale.

The existing infrastructure in the brownfield's case, however, may have legacy systems that are likely to be reused in part or as a whole whenever possible. Reusing different parts of the legacy systems may impede the new system's scalability.

Moreover, legacy systems usually allocate more physical space and are more power-consuming than modern systems, which adds more limitations to the overall system scalability.

Having said that, as we deploy the new system and gradually refresh the existing infrastructure, we will use up-to-date technologies and modern systems that will free up physical space and reduce power requirements. This, in turn, will help us eventually scale the system better.

Technology stack

In a greenfield deployment, we have the liberty to leverage the latest technologies, applications, and tools, which can, among many other things, enhance performance, security, and capability, and prolong the system's overall lifetime.

With the legacy systems in brownfield environments, older technology, hardware, tools, and systems are used, which can introduce limitations on the system's supportability, capabilities, scalability, and future expansions and integration.

Integration

As explained in other aspects of the comparison, in greenfield environments, all the components of the system are new and are designed and built from the beginning to work together seamlessly. Integration, therefore, is not an issue at all.

Integrating a new system with an existing IT infrastructure in the case of a brownfield deployment, however, can be challenging, as the two systems may not be fully compatible. Integration efforts may be needed for old and new components to work together, and even then, the new mixed system may later provide operational challenges that can cause unforeseen system mishaps.

Cost

From an acquisition and CAPEX perspective, building a new system from scratch is higher in cost than building the system from a mix of reused and newly acquired components.

To set up the new system, a certain level of expertise is required that may not be available in-house. The effort and expertise needed to bring the system up and running will certainly have associated costs. However, it could be argued that this cost can be easily offset by the efforts and expertise needed to integrate the new and old components. Possibly different effort and different expertise, but similar cost.

When it comes to OPEX, in a greenfield deployment, we need to consider training costs for the new technologies and systems deployed, as well as potential operational mistakes that can be caused due to the lack of new system hands-on experience. In brownfield deployments, these training costs are usually lower.

Power consumption is typically lower in greenfield deployments, as new systems and technologies are often geared toward power usage optimization.

Another important OPEX consideration is the potential **technical debt** in brownfield implementations. Technical debt is the shortcuts the organization takes to get the system up and running. In other words, this involves taking a band-aid approach to resolving an integration or operational issue during the deployment, and achieving short-term results that can be catastrophic in the long term.

Time-to-market

Time-to-market is an interesting aspect of the deployment and can go both ways. Generally speaking, deploying a system in a greenfield environment takes longer than integrating with an already existing functioning system, as is the case in a brownfield environment. But that would be highly dependent on how complex an existing system may be.

If we are deploying our new system on top of a significantly old or complex disorganized infrastructure, we could argue that deploying a new system fresh from scratch is much more straightforward and a time saver than trying to get both systems integrated successfully.

Risks

This is another aspect of the deployment that can be debated either way.

With a lack of experience with the new infrastructure, new technologies, new systems, new tools, and new applications, system operational mishaps are more likely, and the time to resolution could be higher. In contrast, with no backup system in place, there will be no fallback option if the new system does not perform as expected.

But again, if the old system in the brownfield environment is too complex or disorganized, the risks would be higher in a brownfield deployment due to complexities in integration, potential technical debt, old unsupported components, and more.

Staff onboarding

Companies that are already using an existing system have a better understanding of how it works and insights into its operations and potential issues, which can make the deployment process and system operations smoother.

In greenfield deployments, training and accumulated experience are needed before staff can start to become familiar with the system details.

User adoption

User adoption of the new system may require adapting to new ways of performing day-to-day tasks, a shift in internal and potentially external business operations, and how the organization deals with internal and external customers. This shift may require a change in organizational culture, which can pose a significant challenge to the successful implementation of the system and reveal operational shortcomings after deployment.

In a brownfield environment, the updated system capabilities could be incremental or somewhat transparent to the user, which makes user adoption much easier and faster compared to deploying a completely new system, as in the case of a greenfield deployment. A successful and gradual user adoption helps uncover potential design, implementation, and operational deficiencies that can be quickly addressed and fixed.

In either case, user training is needed, but certainly, in a greenfield, training is more complex and involved over the brownfield case.

The following table summarizes the comparison between greenfield and brownfield aspects of the deployment:

	Greenfield	**Brownfield**
Flexibility	High	Low
Scalability	High	Low
Technology Stack	Flexible and optimized	Restrictive
Integration	Minimal to none	High efforts
Cost	CAPEX is high but better OPEX	CAPEX is lower but higher OPEX
Time-To-Market	Usually longer	Usually shorter
Risks	Usually higher	Usually lower
Staff Onboarding	Longer process	Shorter time
User Adoption	Slow	Fast

Table 11.2: Greenfield versus brownfield

In this section, we discussed the main differences between greenfield and brownfield deployments, the pros and cons of each environment, what to consider, and why.

In the next section, we will go over how we can overcome deployment challenges in both environments and the deployment best practices in each case.

Overcoming deployment challenges

We now understand the different aspects of system deployment in greenfield and brownfield environments, as well as the several challenges that can be presented during the system design and implementation. In this section, we will cover some of the concepts, strategies, and approaches for mitigating these challenges to ensure a smooth and successful system deployment.

We should begin our deployment project with a solid project team that possesses diverse skills and experiences in technical areas of the project, deployment and project management, and vendor management.

In the absence of in-house experience, outsourcing one or more project experience areas through third parties may be necessary. Partners may include system integrators, and/or equipment vendors, and value-added resellers.

This experienced team will help conduct thorough planning and research to be able to understand the specific needs of the organization, potential risks, local regulations, and any necessary compliance needs.

Compliance with industry and local regulations is an essential part of the project. Aside from the technical aspects and technologies of the project, a system processing credit cards, for example, will require team members who are experienced in PCI compliance and rules. A healthcare system deployed in the United States, for example, may need members who have experience with HIPAA compliance needs, and so on.

Project management is key to establishing clear team communication and collaboration. The project managers help track the project process, changes, and requirements, and ensure that the timelines and goals are met throughout the project cycles. The project managers also make certain that all stakeholders are properly informed and engaged throughout the different phases of the project.

The type of project management style or approach is dependent on the organization itself, the timeline, and the implementation details and technologies deployed. Whether it is waterfall, agile, scrum, or something else, the project manager must decide with the team.

Addressing deployment challenges is a task that is pursued within the project cycle. Our focus here is on this aspect of the project cycle, particularly concerning greenfield and brownfield deployments.

Before any deployment activities, complete visibility of the deployment risks is necessary. Therefore, it is imperative to develop a clear deployment risk plan to be able to identify the risks and mitigate each risk to ensure a successful deployment.

The following chart illustrates the risk management process for our deployment cycle. The process should start with identifying the risks, determining ways to avoid or minimize them, developing a mitigation plan, and continually testing, monitoring, and reviewing the deployment to update the mitigation plan for any new risks or challenges that arise during implementation:

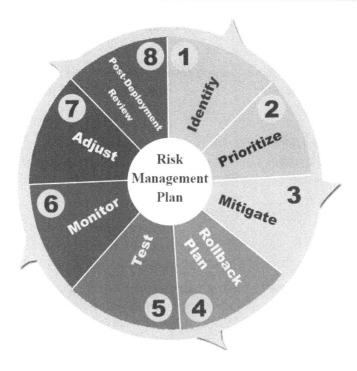

Figure 11.1: Main steps of overcoming deployment challenges

In the following subsections, we will go over the main activities of a risk management plan and how each phase of the plan is relevant to our deployment activities.

Identify deployment risks

To address the deployment challenges, we start by identifying the potential risks we may encounter when deploying our system. In a greenfield environment, the risks we identified earlier are as follows. The **greenfield risk** is referred to as **GR**:

- **GR1**: High CAPEX
- **GR2**: Deployment time
- **GR3**: System failures due to lack of training and experienced staff
- **GR4**: Slow or lack of user adoption

In a brownfield environment, the risks we identified earlier are as follows. The **brownfield risk** is referred to as **BR**:

- **BR1**: System capabilities limitation due to potentially low flexibility and scalability, and some reused legacy technologies

- **BR2**: High OPEX

Now that we have identified the potential risks, it's time to prioritize them based on their likelihood and potential impact on the project. This will help us calculate the **risk exposure** and plan and allocate proper resources effectively.

Prioritize risks

Risk exposure is the risk probability multiplied by the impact of that risk on the deployment project. The higher the risk exposure, the higher the priority of working on mitigating that risk should be.

Figure 11.2 shows a color-coded risk exposure matrix for the previously identified risks. Please note that the level of risk exposure can greatly vary between organizations based on various factors, such as project complexity and organizational needs, the stability and complexity of the existing system, project requirements, budget, timeline, and more:

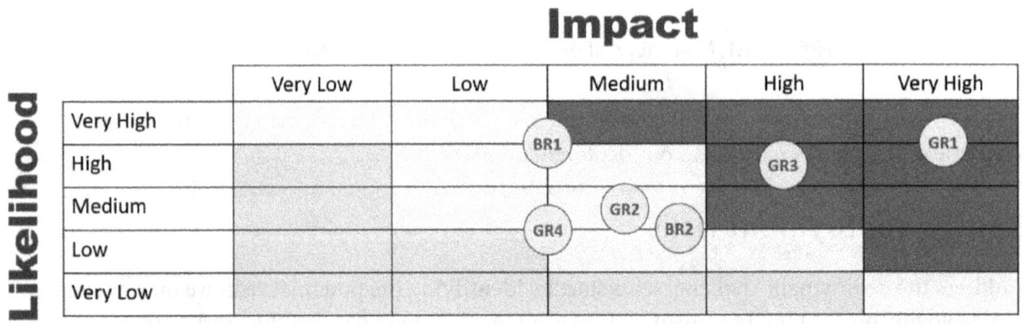

Figure 11.2: Risk exposure matrix

We always prioritize from the top right to the bottom left of the risk mitigation chart. The red zone, where the risk and the likelihood are Medium to High, is where we need to start allocating resources. This is followed by the yellow zone, then the green zone.

So, we should start with greenfield risk #1 (GR1), how to mitigate high CAPEX risks, then GR3, where we lack experienced staff to deploy the new system, then GR2, where the new system deployment time may be an issue, and then conclude the mitigation of greenfield risks by addressing GR4, where we may face a slow user adoption of the new system.

For brownfield risk mitigation, we start with brownfield risk #1 (BR1) since it has a higher exposure in the red zone of the matrix, then do BR2.

Risks are not always avoidable. Risks often aren't. In cases where risk elimination is not possible, the mitigation plan has to address how we can at least minimize the risks to a manageable level.

Many organizations choose to ignore risks with low/very low likelihood and low/very low impact. The process of mitigating risks in this particular case may very well be costly and riskier than the risk itself.

Developing and implementing a risk mitigation plan

We need to develop a mitigation plan to manage the identified risks. This plan should include specific actions that will be taken to mitigate or eliminate the risks, resources involved, and contingencies in case the risks do occur.

We also need to identify our risk mitigation strategies based on the calculated exposure of each risk. As we will see in the next few subsections, these strategies can include implementing operational safeguards, extra measures using additional system components, testing the system before deployment, training both system users and administrators to ensure they can effectively use the new system, and developing a rollback plan in case of any unexpected deployment issues.

Let's apply all these to our deployment project's identified risks.

GR1 – high CAPEX risk

When addressing the high CAPEX risk, we need to focus on a few things – first, the project budget, second, the system and project requirements, and third, how to use effective negotiation skills to acquire the necessary infrastructure that would successfully fulfill all of the desired system requirements.

Sometimes, of course, budget and time constraints become a barrier to acquiring all of the requirements' wish list items. Therefore, it is important to prioritize your requirements, especially if you have strong budget constraints.

The objective is to acquire the infrastructure that would get us all of our wish list requirements; however, at some point, we may need to give up some of the nice-to-have features of the system for the sake of meeting our budget. This is where our negotiation skills become vital. The stronger our negotiation skills, the greater the likelihood of successfully deploying the new system and meeting all the requirements within budget.

It is useful to go over a few important negotiation techniques that can help us achieve our objective.

Start by conducting thorough research on all vendors that can be involved in providing the infrastructure assets. This includes the technical and business strengths and weaknesses of each potential vendor, their list prices, the quality level of their deployment and operational support, their product roadmap, and their future business outlook.

Then, we can come up with multiple options from our vendor research. Having multiple options gives us more flexibility and bargaining power during our vendor negotiation phase.

Out of all the available options, having a clear **Best Alternative To a Negotiated Agreement (BATNA)** is critical. BATNA means having the most favorable option we can achieve if the negotiations fail and no agreement is reached with the infrastructure vendors. It is the fallback option that we can rely on if the negotiations do not go as we initially planned with a specific vendor.

Having a clear and well-defined BATNA is important in displaying strong bargaining power during the negotiation process. A BATNA gives the vendors a sense of how much they can miss out if they do not close the infrastructure acquisition deal.

BATNA also helps us understand when it's time to walk away from the negotiations if the vendors are not willing to meet our infrastructure needs with the available budget. It is about the alternatives that are available to us and the consequences of choosing or not choosing a specific vendor.

Larger and highly reputable organizations can use leverage. Leverage is the ability to use your reputation, market position, or market size to influence the outcome of the negotiation. Leverage can be very effective in striking a good deal with vendors. Vendors generally strive to gain a foothold in large organizations to create a customer reference or as a way to build more business out of that particular deal.

Having said all that, it is essential to show a certain level of flexibility during the negotiation. Flexibility will demonstrate our willingness to compromise and is likely to lead to a strong long-term and healthy business relationship with the vendors:

Figure 11.3: Negotiation strategies

Figure 11.3 sums up the negotiation strategies you could follow to minimize budget overrun risks due to infrastructure cost and the overall project CAPEX.

Another effective way of mitigating CAPEX risks is to use a cloud provider in deploying your new infrastructure, especially during the initial testing and adjustment phases of the project.

Using containerization and virtualization to build new workloads is another way to help reduce CAPEX risks. However, by doing so, we may introduce some OPEX and other types of deployment risks, especially if the staff is not trained enough on cloud environment deployments.

GR2 – deployment time risks

When deploying a new project, especially a project that involves new technologies, techniques, and concepts, the potential of running behind schedule cannot be ignored. There is always a learning curve associated with the overall project cycle.

Moreover, potential frequent changes and scope creeps are also important to account for. This emphasizes the importance of engaging a skilled team from the start even more.

A highly skilled and well-trained project team will help minimize scope creep and time delay risks. In the absence of in-house subject matter experts, hiring an external party as a system integrator and building an adequate training plan for the team become critical when managing time delay risks.

There may be factors other than team skills that can contribute to project timeline and scope creep risks. As part of our risk mitigation plan, we need to account for all of these factors and make sure they are controlled to consequently control their associated risks.

MSA systems can be very complex and burdensome to deploy. The more complex the system is, the more difficult it can be to predict and manage the deployment timeline.

In brownfields, the deployment of the new system often depends on the availability and operational stability of the already existing systems. Furthermore, integrating the new system with the existing systems can add another level of complexity and time-consuming tasks, all of which can potentially prolong the overall deployment timeline.

One way to mitigate the effect of complexity on the overall risk is to avoid forklift and big bang changes. Rather, use a trickle approach by breaking down the deployment into multiple simple phases and stages. Follow the famous Einstein rule, *"Everything should be made as simple as possible, but not simpler."* Simplify as much as possible, but not in a way to compromise the system's functionality or reliability.

Although thorough testing is necessary to ensure the new system's reliability and adherence to the organization's needs, over-analyzing and over-testing often happen in complex projects. This process can significantly delay the system deployment.

One way to address the testing delays factor is to consider running the system in production in a **Limited Availability (LA)** fashion, a beta version, or a pre-launch period of some sort. This LA approach will help us apply real user traffic to the system while we focus on monitoring and making system changes as needed before we transition the system into full-scale production.

GR3 and GR4 – system failure risks and user adoption risks

User adoption can highly depend on any new **user interface/user experience** (**UI/UX**) changes or added complexities. Assuming a well-designed system's UI/UX, system failure and slow user adoption risks are also dependent on the experience of the project and operation teams, and how familiar the users are with working on the system.

To normalize the experience and user familiarity risks, it's important to first include a solid UI/UX design team from the outset of the project, then adopt a top-down approach by securing a strong buy-in from project sponsors and executives.

The top-down approach will help adopt the organization's processes and changes necessary to create a cultural shift in conducting the business using the new system. The buy-in can also help enforce a training program for both system users and the project team. This training can significantly help in bridging the gaps between the existing experience of the project team and the experience needed.

BR1 – system capability limitations

Because we are deploying the new system with some older components being used in the existing system, we are likely to run into integration incompatibilities, limitations in the older hardware and software features, and limitations in scaling in terms of traffic, data loads, or storage.

Additionally, a system with legacy components may become outdated sooner than anticipated. This can result in obsolete software and hardware components that are no longer usable or valuable, or in vendor out-of-support announcements, which would shorten the life span of the new system.

When system components are out of support, vendors can no longer provide part replacements or software updates, or even assist in case any operational issues arise on the out-of-support components. This can severely impact the system's reliability and jeopardize the organization's business continuity.

To mitigate the risk of system capability limitations, we must have clear visibility and understanding of the vendor product map for each reused legacy system component and clear visibility of the component's dependencies. This understanding shall help us assess the impact of that component on the scalability levels, and future operational reliability and stability of the system being deployed.

BR2 – high OPEX

As we have previously explained, integrating both new and older systems can introduce many deployment and operational challenges that can make the system much more complex, and introduce technical debt, high maintenance costs, and high operational costs to keep the system running smoothly.

Security is also a major concern in brownfield deployments. Technical debt, along with mixing older and new components, may introduce vulnerabilities that were not present in the existing infrastructure, which can lead to costly data corruption, data loss, data recovery, security breaches, and irreversible reputational damage.

To mitigate these risks, organizations should thoroughly evaluate the impact and have a clear cost-benefit analysis of each reusable component of the existing system before proceeding with the deployment.

Furthermore, having a robust disaster recovery and backup plan in place in case of data loss or corruption is key to mitigating some OPEX risks.

All the preceding risk examples and their mitigation strategies should be thoroughly discussed as part of the developed risk mitigation plan, with step-by-step guides and documentation.

Should any of these risks take place, especially in the case of brownfields, where an existing system may already be running, a comprehensive rollback plan should be executed immediately. The rollback plan is the next step in how to overcome the deployment challenges. The following subsection takes us through what a rollback plan is and what it entails.

The rollback plan

Remember Murphy's Law, which states *"Anything that can go wrong will go wrong"*?

How often do we create a solid and well-crafted implementation or migration plan, expecting a smooth change in the system, but then experience unexpected and bizarre behavior during the execution of the plan?

A rule of thumb here is that anything can go wrong during the deployment. We put enough planning and precautions for nothing to go wrong, but unfortunately, things do not always work in our favor. We may still overlook things, system bugs may get triggered, equipment mishaps may happen, and so on.

Therefore, developing a rollback plan is necessary to be able to maintain business continuity. We should have the plan built in a way that it includes clear steps and procedures to move back to the initial system state before the change and resume normal operations quickly.

Adopting a phased deployment approach, as previously discussed, helps to quickly roll back only a portion of the change, which helps us avoid wasting resources, valuable change window time, and efforts invested during the deployment, and in developing the deployment plan.

Test, monitor, and adjust

The next step in overcoming deployment challenges is to test and validate the system's requirements and functionality to ensure it meets the performance, functional, and operational requirements.

The project teams, when under time pressures, often deprioritize system security over system functionality. Data security and data protection, when overlooked, can severely impact the project's overall deployment and reliability, especially with systems that handle critical user data and have to comply with certain regulations and compliance acts, as in the case of PCI or HIPAA.

Therefore, the test plan has to have a specific section dedicated to system security and compliance testing. Hiring a specialized firm in the area of security and compliance helps minimize the risk of data breaches or other security incidents.

As we gradually apply test and production load to the system, the test plan should be able to ensure the systems being deployed are scalable and flexible enough to adapt to changing needs and requirements. This is a critical aspect of the test plan during a brownfield deployment since the integration with the existing system may hinder the overall system's capabilities.

Then, we need to continuously and carefully monitor and review the identified risks throughout the project to ensure that the risk mitigation plan is effective and that any new risks are identified and addressed promptly.

Any newly identified risks will have to be included in the mitigation plan through the project change management process. The newly identified risks and mitigation strategies will need to be communicated to all stakeholders, including the project team, the organization's management, and the system users.

Post-deployment and pre-production review

Once the deployment is over, the system is operational, tested, and running in a pre-production or LA manner. Just before closing the deployment project, we need to evaluate the effectiveness of our risk management plan, identify any areas for improvement, and document the outcome of our findings. The outcome can then be integrated into the same project deployment effort or conclude the existing deployment and initiate a new project for that matter.

The post-deployment assessment will ensure the new system's continued stability, performance, and reliability.

In brownfields, we may end up running in a bi-modal approach, where both systems are running at the same time and serving users at the same time but in different ways and at different levels. In this case, we need to consider building specific roles and responsibilities matrices for each system. This helps streamline operations and increase the system's supportability.

Summary

In this chapter, we covered greenfield and brownfield deployments, the difference between each, their pros and cons, risk details during the deployment process in general, and the specifics of each risk in each deployment case.

We also provided examples of the risks associated with greenfield and brownfield deployments, along with the strategies to mitigate these risks, to gain a better understanding of the challenges involved in successfully deploying a new system.

The topics that were discussed in this chapter act as introductions to what we will be learning in the following chapter. In the next chapter, we will apply some of what we have learned in this chapter and discuss ways to test, monitor, and update our new ABC-Intelligent MSA system.

12

Deploying, Testing, and Operating an Intelligent MSA Enterprise System

In the previous chapters, we talked in detail about microservices, monolithic architecture, the pros and cons of each architecture, how to transition into MSA, and how to make the MSA system smarter using AI services. We also discussed, in *Chapter 11*, some of the best practices for deploying the MSA system.

In this final chapter, we will integrate all the topics and concepts covered throughout the book to understand how we can apply what we have learned through hands-on and practical examples.

Before we dive into the details, we need to understand what existing system we have in place first.

Obviously, every organization is different and has different deployment needs, criteria, and dependencies. Some organizations will deploy in a greenfield, and others in a brownfield. In order to walk you through detailed practical examples and steps for deploying, testing, and operating an intelligent MSA system, we will assume a brownfield environment with an existing monolithic architecture system.

We will sometimes use our ABC-Monolith as an example of the existing system to illustrate the concepts covered in the chapter. In this chapter, we will cover, the following topics:

- Overcoming deployment dependencies
- Deploying the MSA system
- Testing and tuning the MSA system
- The post-deployment review

Overcoming system dependencies

Before deploying the ABC-Intelligent-MSA system we built earlier in *Chapter 10*, it is important to first decide what our deployment strategy should be. Based on the requirements, cost, complexity, and pros and cons of the deployment strategies we discussed in *Chapter 11*, we believe the best deployment strategy for our ABC system would be a mix between the ramped deployment and canary deployment strategies.

This deployment strategy will allow us to keep the ABC-Monolith system online and serve users uninterrupted while we deploy the new ABC-Intelligent-MSA system. We will gradually replace older components in ABC-Monolith with the corresponding microservices in the ABC-Intelligent-MSA system. This can be accomplished by routing traffic from the older components to those ABC-Intelligent-MSA system microservices.

Although this trickle approach has lower cost, lower complexity, and lower risk than other deployment approaches, we still need to carefully study the incompatibilities, dependencies, and the proper integration between older and newer components.

Furthermore, we will need to evaluate which of our infrastructure and existing system ABC-Monolith's components can be reused in the new architecture, if any.

Reusable ABC-Monolith components and dependencies

We cannot think of a specific ABC-Monolith code base component that can be reused as is without modification. All of the ABC-Monolith components will have to be either rewritten from scratch or modified to different degrees to be compatible with the ABC-Intelligent-MSA system.

Some of the ABC-Monolith and existing infrastructure components that we know can be reused are the business logic itself, server infrastructure, operating systems, virtualization infrastructure, data storage, network infrastructure, existing monitoring, and network management tools, and some of the software and database licensing. Nevertheless, even these components may need to be updated or upgraded in order to perform the functions of the new system.

In our system installation, command line, and code examples listed in earlier chapters, we had the most updated Ubuntu, Python, and database versions. In a real-life situation, however, that may not be the case; we will likely have the monolithic application running in an older operating system and have an older Python and/or database version.

These situations may produce some incompatibilities between the older components and the newer ones. An older Python version, for example, may have some deprecated functions that are no longer valid with the new MSA code base, and hence, will require some updates or upgrades to the existing system. Furthermore, the potentially different technology stack may also produce more dependencies.

In order to minimize these dependencies, we would rather deploy the new system components on a separate server or virtual infrastructure with their own environment, including their own data storage and using their own technology stack. The new environment will have a container engine that will carry all of our ABC-Intelligent-MSA microservices.

It is important to note that each system is different, and the specific reusable and non-reusable components will vary depending on the existing monolithic system. A thorough analysis and evaluation of the existing system's components is necessary to determine what can and cannot be reused when migrating to a microservices architecture.

Mitigating ABC-Intelligent-MSA deployment risks

Some of the risks discussed in *Chapter 11* are relevant to our scenario. However, we still need to determine which CAPEX risks are applicable, examine the risks related to deployment time, potential service disruption, and OPEX, and take specific actions to mitigate these risks.

Since we are using containers on top of virtualized infrastructure in our implementation, CAPEX risks are significantly reduced. As long as the existing infrastructure has the storage and workload capacity to absorb the new ABC system, we are safe. If additional infrastructure resources are needed, we may then need to look into some capacity planning and upgrades to be able to run the system during and after the deployment.

Adopting a trickle migration approach gives us the chance to catch up quickly with any learning curve involved with the new technologies being deployed, which, in return, helps mitigate system failure risks and deployment delays.

The ramped deployment strategy also helps mitigate other OPEX risks. As we will discuss in this chapter, during the deployment, we can test and monitor the performance of the newly deployed components, identify and resolve any issues, and make necessary adjustments before redirecting all traffic to these new components.

Another way of mitigating OPEX risks is to establish a robust change management process by establishing a structured and transparent process for managing changes. This includes creating clear guidelines for how changes will be proposed, evaluated, approved, and implemented, as well as communicating the changes to relevant stakeholders.

Part of the change management process is the rollback plan. The rollback plan is essential to bringing back the system to an operational state if a specific technical change is unsuccessful. The following are the steps we need to consider to build a successful rollback plan:

1. Specify some checkpoints for the change where a rollback may be needed. In our example, and should the ACL be used, the ACL would be deployed prior to switching any traffic to the new microservice. During the change (and right after switching some test traffic to the ABC-Intelligent-MSA), a few good checkpoint examples would be as follows:

   I.   Testing the payment verification communication between ABC-Monolith and the ACL

   II.   Testing how the ACL processes the requests

   III. Testing the communication between the ACL and the ABC-Intelligent-MSA system

   IV.   Testing how the overall end-to-end requests are handled and whether they are processed as expected

   Common Docker and Linux commands to test and troubleshoot the communication between the ACL, the monolith, and the MSA include the following:

   - `curl`, to simulate an API call to the ACL or a specific microservice

   - `netstat`, to check whether a specific service is actively listening to connections, what the listening port is, and whether there are any active connections

   - `docker inspect`, to return detailed JSON information about a specific microservice's configuration, state, and network settings

   - `docker log`, to view the logs of a running container

2. Develop a plan for reversing the change at each of the preceding specified checkpoints.

3. Whenever possible, test the rollback plan in a test or staging environment to ensure it is workable and complete.

4. Know the time by which the rollback plan needs to be completely executed at each of the specified checkpoints, and allocate a reasonable amount of time in your change for it.

5. Monitor the system throughout the change and make adjustments as needed.

6. Have a post-mortem after the change, especially in case of a change failure.

7. In case the rollback plan is executed, the team will then need to include in the post-mortem the reasons for the change failure, and how effective the rollback plan was. They need to accordingly make the necessary adjustments to the deployment and rollback plan before scheduling another change.

By this point, we should have a clear understanding of the deployment dependencies and risks and be able to determine methods for mitigating them. We are ready now to create a deployment plan and execute it in a manner that minimizes downtime and maintains business continuity.

In the next section, we will build the ABC-Intelligent-MSA system's deployment plan in the presence of the running ABC-Monolith system.

Deploying the MSA system

In *Chapter 9* and *Chapter 10*, we discussed in detail how to install Docker, containers, and other components for our ABC-Intelligent-MSA system. This installation was mostly done in a lab environment with no specific regard to any existing system in the environment. We were basically just simulating a real-life development or staging environment.

In this section, we will focus rather on how we can take the ABC-Intelligent-MSA system we built, and gradually migrate it into a brownfield production environment where we have the ABC-Monolith system already running in production. The goal is to ramp up the ABC-Intelligent-MSA system's operations until the system is able to carry the entire existing traffic, then completely phase out the old ABC-Monolith. Everything should be done with minimal operational interruptions.

The current status by now is that we still have ABC-Monolith running in production, and the ABC-Intelligent-MSA running in the staging environment. The following are detailed broken-down deployment plans with their execution steps.

The anti-corruption layer

Both our ABC-Monolith and ABC-Intelligent-MSA systems use the same type of RESTful APIs and the same JSON data formats. Moreover, our demo system is not complicated enough to justify an ACL. We, therefore, won't be needing an ACL in our migration. However, we developed an ACL in our demo just in case you decide to try it out.

In case you are interested in trying the ACL, the first step you would need to do is to get the ACL up and running. The ACL will act as a buffer and handle the communication between the ABC-Monolith and the ABC-Intelligent-MSA systems.

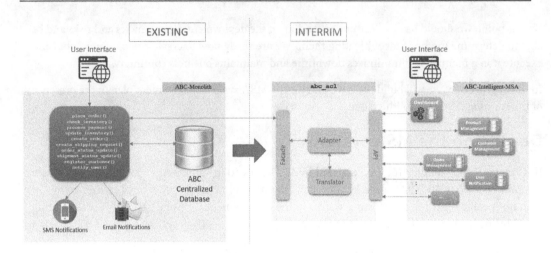

Figure 12.1: Deploying with the ACL

ACLs are usually a specific custom-built code for the organization's specific situation, the old system, and the new MSA system. We built the `abc_acl` ACL for our ABC system. The `abc_acl` code can be found in our GitHub repository.

It would make much more sense to deploy all the new components, including `abc_acl`, on a separate host or virtual workload. In our lab examples, however, and for simplicity, we are building the new system's containers on the same host that's running the ABC-Monolith.

We built the **Facade**, **Adaptor**, and **Translator** components all together as part of the ACL in one microservice. The Facade is created to interface with the ABC-Monolith, the Adaptor to interface with the ABC-Intelligent-MSA, and the translator for input/output data format mappings. Since we are using the same data formats in both the monolith and the MSA systems, the translator code is not doing any processing and is just used as a placeholder.

We can set up and start the `abc_acl` microservice the same way we did with other microservices in *Chapter 9* and *Chapter 10*, using the `docker build` command to build the `abc_acl_image` image from the Dockerfile, then using the `docker run` command to create the `abc_acl_container` container, as follows:

```
$ docker build -t abc_msa_customer_management ~/
```

Once the image is successfully created, use the following command to run the container, and start listening to port TCP/8020 on the host's IP:

```
$ docker run -itd -p 8020:8080 --name abc_acl_container abc_
acl_image
```

Now the ACL is running, it is time to test it before routing any traffic to it. We can do that using the shell `curl` command as we did in the previous chapters, or we can use some of the ACL built-in API tools created to verify the connection.

The following is a `curl` command issued on the host machine to ensure that the ACL is running successfully:

```
$ curl http://192.168.1.100:8020/
<!DOCTYPE html>
<head>
    <title>The Anti-Corruption Layer Microservice</title>
</head>
<body>
    <p>This is the ACL Microservice Part of ABC System. This ACL
is used as part of the process of migrating ABC-Monolith to the
new system, ABC-Intelligent-MSA </p>
</body>
```

The following is an example of another way to test the ACL – more specifically, to test the communication between the ACL and both ABC-Monolith and ABC-Intelligent-MSA systems:

```
$ curl http://192.168.1.100:8020/api?func=test_com
{"monolith_com_test": "Communication SUCCESSFUL", "msa_com_
test": "Communication SUCCESSFUL"}
```

The first `curl` command ensures that the ACL is listening to API calls from both ABC-Monolith and ABC-Intelligent-MSA, while the second `curl` command ensures that the ACL can successfully communicate with both systems.

The ACL operation is now verified; in the next subsection, we will start migrating the MSA services from the staging (or lab) environment to the actual production environment running the old monolithic system.

Integrating the MSA system's services

With the ACL now up and running and tested successfully, we are ready to start switching specific traffic to specific parts of the ABC-Intelligent-MSA. We will, however, use direct interaction between both the monolith and the MSA system since the ACL is not really needed in our demo example.

Please note that, depending on the existing monolith structure, design, and system's code base, this process could either be very straightforward or as complicated as can be. Our deployment strategy requires some code changes in the monolith system to be able to route some parts of the traffic to the new MSA.

For that reason, we may very well choose, in some systems, to have the MSA completely tested in a staging environment, then put the MSA on an LA period where partial production traffic is passing through the system for deeper testing. Then, once comfortable with the new MSA system's performance, we can just start forwarding the entire production traffic, and finally, shut down the old monolith.

Figure 12.2 shows a high-level view of the migration before and after status. During the migration, we will route a specific function of ABC-Monolith to one microservice in ABC-Intelligent-MSA. That microservice should be able to replace the corresponding function in the ABC-Monolith system. After we test the operation of that part of the migration, we then move traffic of another monolithic function, then another, and so on, until we end up migrating all of the ABC-Monolith functions to the ABC-Intelligent-MSA system.

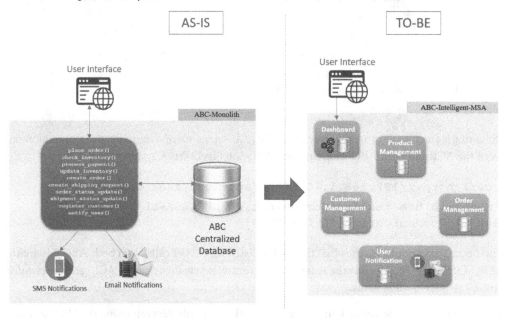

Figure 12.2: A high-level view of where we are and where to be

We can start with a simple microservice such as **Notification Management** (abc_msa_notify_ user_container). We can route the traffic destined to the notify_user() function in the ABC-Monolith by replacing the function's code with an API call to abc_msa_notify_user_ container. All user traffic will still flow through the ABC-Monolith, but all user notifications will be processed through the ABC-Intelligent-MSA.

In the same manner, the **Customer Management** microservice (abc_msa_customer_management_container) should replace the register_customer() function from the ABC-Monolith, and **Order Management** should replace place_order() and order_status_update() functions, and so on.

As the system stabilizes, we gradually migrate to other MSA services. That migration cycle is shown in *Figure 12.3*.

By following the migration cycle, eventually, all of the ABC-Monolith functions will be replaced with microservices in the ABC-Intelligent-MSA system.

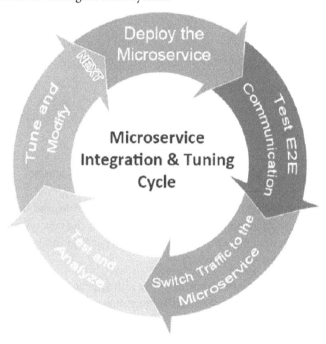

Figure 12.3: The microservices integration testing and tuning cycle

Figure 12.4 shows a snapshot of the system status during the migration process. In the figure, we have the **Notification Management**, **Customer Management**, and **Order Management** microservices successfully migrated, but not any other microservice yet.

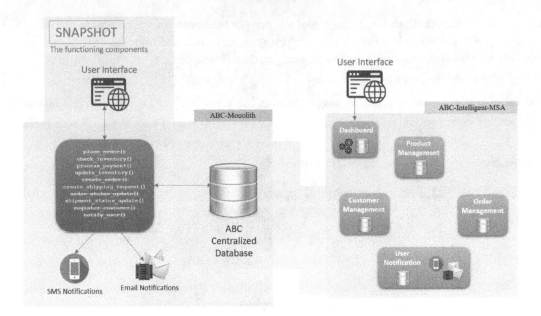

Figure 12.4: A system snapshot during the migration process

Monitor the system closely during the monolithic function migration, and use the rollback plan if necessary. Once the last ABC-Monolith function is migrated to the new system, we will need to carefully run an end-to-end test on the ABC-Intelligent-MSA system to ensure the system is running properly and independent of the ABC-Monolith.

Testing all microservice logs and stats is essential in the testing process. We need to have a formal testing process in place every step of the way during the migration process. The test process is described in more detail in the next section.

Keep both systems running for a period of time just in case some overseen issues take place and always be prepared with a contingency plan.

The final step is shutting down the monolith. If the migration steps were followed and tested correctly, user traffic and system operations should not be impacted. However, complex systems may have a component or more still processing traffic. To avoid business interruptions in this situation, it is best to shut down the monolith during a maintenance window to allow the migration team to analyze any unforeseen issues and create a plan to resolve them.

In this section, using our ABC system, we explained the MSA system deployment process using an ACL and using a direct monolith-to-microservices approach. We covered the steps to be taken, what to watch for, and how to make the transition to the new system as smooth as possible with minimal system interruption.

In the next section, we will cover the formal test methodology that should be planned and followed after every microservice migration to ensure system reliability and stability.

Testing and tuning the MSA system

Prior to deploying microservices, a formal testing or QA process should be applied to each microservice to prevent errors during deployment and in production.

There are a couple of tests that need to be performed on the MSA system microservices before deploying them in the production environment. First, testing the microservice itself as a standalone and before integrating it into any parts of the ABC system – what we refer to it as, **unit testing**. Second, testing the integration of that microservice into the ABC-Intelligent-MSA system – what we refer to it as **integration testing**. And third, testing how the microservice functions during an interim mix of operations between the ABC-Monolith and the ABC-Intelligent-MSA systems.

Testing the ABC system functions every time a new microservice is deployed is crucial to ensure a successful migration and that the system is able to properly function and sustain the applied traffic load and user requests.

Building structured test cases is an important part of the testing process. Test cases are a set of steps that describe how to test a specific feature or functionality of a system. These test cases should be well defined, easy to understand, and should cover all possible scenarios.

Creating a test case should include the following main steps:

1. Identify the requirements of the system and the feature or functionality that we want to test.
2. Write a test case that describes the steps to be taken to test that feature or functionality.
3. In the test case, specify any prerequisites that are required to run the test.
4. Identify the pass/fail criteria based on the expected outcome of the test case.
5. Run the test case and compare the test results to the expected outcome. Accordingly, and based on the pass/criteria specified, record the result of the test in simple PASS or FAIL terms.

The following is a simple example of a test case for the `Notification Management` microservice. The test case verifies that the microservice is actually sending an SMS notification to a registered user's mobile number. Another test case should also be written to test the microservice's email *Send* functionality. We can write as many test cases as needed for each individual microservice, and for the system functionality overall.

Test Case Details	
Title	ABC-Intelligent-MSA Notification Management Microservice SMS Functionality
ID	002912
Description	To ensure that the microservice is actually sending an SMS notification to the specified mobile number
Requirement(s)	Access to the SMS gateway.
Test Setup	Have access to the receiver test phone +1 (555) 555-5555. Have access to the Ubuntu test environment. Verify SMS gateway access.

Procedure	Ensure the `abc_msa_notify_user_container` container is running, or start it as follows:
	
	`docker container start abc_msa_notify_user_container`
	
	Issue a shell `curl` command as follows:
	
	`curl http://192.168.1.100:8010/api?func=send_sms&num=15555555555&msg=order+received`

Test Type	Unit testing
Pass/Fail Criteria	Test case passes if you receive the message "order received" on the test phone

Table 12.1 – A sample test case for notification management in ABC-Intelligent-MSA

Testing the AI services of the ABC-Intelligent-MSA system can be more challenging, and the conventional test case approach may not be sufficient. Testing the AI part of the system will require a multi-level approach that would require including the microservice itself in isolation (unit testing), integration testing, functional testing, performance testing, data validation testing, and human-in-the-loop testing. By using all these approaches together in building your test cases, we can ensure that the AI components of the system are functioning as intended and are making accurate predictions and decisions.

In this section, we covered the importance of building a structured testing process and built a test case example as part of our system's testing process. We discussed how to create a test case and identify the requirements and expected outcomes.

In the next section, we will talk about the importance of conducting a post-deployment review after the completion of the ABC-Intelligent-MSA system deployment. The section will also cover the different types of post-deployment reviews, including user feedback reviews.

The post-deployment review

The ABC-Intelligent-MSA system is currently running, but it hasn't been operational for a sufficient amount of time to guarantee its stability and resilience under typical traffic patterns and loads. A post-deployment review is crucial for ensuring the success of the ABC system deployment and its compliance, as well as enhancing its functionality and overall user satisfaction.

During the post-deployment review, we will need to monitor the system closely and look for any errors, bugs, or any other operational problems that may happen. Then, we will need to make recommendations for addressing system issues and making necessary improvements to the system to ensure that the system is meeting the user requirements it was created for.

We need to have special monitoring for the AI services we built in the system to make sure they are performing as they are supposed to and continuously improving themselves and the system's operations overall. A closer look at the AI services logs that we discussed in *Chapter 10* is important to ensure the system's stability and enhanced performance.

The following are some of the aspects and criteria that need to be considered when conducting a post-deployment review.

Checking the new system's performance

We start by defining performance metrics, which will help us create a baseline for what to expect from the system, in terms of response time, user interactions, network traffic, and so on. We can use tools available on the internet or the `ms_perfmon.py` we previously discussed in *Chapter 10* to measure the performance of the new system and compare that to the monolith's performance.

The variance between both the old and the new system's performances would highly depend on the design, architecture, operational criteria, and infrastructure used in both cases.

Identifying and fixing system defects

This goes back to the testing and tuning process discussed earlier, and how the process should be conducted. It is important to point out here that post-deployment, identifying system defects is not yet part of the QA process until they are first documented in the organization's defect tracking system.

We are talking here about monitoring the operational aspects of the system and ensuring proper system supportability. The support process may very well lead to filing specific issues found in the system post-deployment. Later, a thorough investigation of customer support cases with their severity levels will need to be conducted to address and fix these issues.

System issues can also be identified by gathering customer feedback, as we will discuss in the next couple of sections, as well as from the outcomes of the different audit processes conducted on the system post-deployment.

Compliance

Perform regular maintenance and updates to the system to keep it running smoothly. As briefly discussed in *Chapter 8*, a considerable part of compliance can be done through automation or commercial tools. The tools will help audit the system for different types of compliances, such as the GDPR, PCI, HIPAA, SCSEM, and so on. The specific compliances that an organization has to comply with will depend on the organization's business itself, the nature of the system, and what processes and users it is serving.

Start by identifying the relevant regulations and standards that apply to the new system. This may include data privacy regulations, industry-specific standards, and cybersecurity standards.

Conduct a risk assessment the way we described in the previous chapter, to identify any potential areas of non-compliance and their associated risks. This may involve reviewing the process of the system's design, architecture, and data processing. Then, put together a mitigation plan to mitigate the identified risks.

Make sure the organization's staff are fully aware of compliance, its importance, and the individual roles and responsibilities in that regard. Keeping the staff trained is another aspect of keeping the organization compliant with specific rules, regulations, and specific industry compliances.

The compliance process is not a one-time thing, the organization has to conduct regular audits to maintain that compliance. Audits may include regularly running specific automated audit tools, and conducting manual system audits by checking system logs, data checks, physical and digital security checks, and so on.

System maintenance and updates

Just like your preventive car maintenance, performing regular system maintenance and updates is important to keep the system running smoothly with no sudden unplanned failures.

By planning, preparing, testing, implementing, monitoring, documenting the process, and taking a proactive role, we can ensure that our newly deployed system is functioning as expected and able to minimize operational interruptions. The following are a few points to consider in the maintenance plan:

1. Put together a regular maintenance plan. This is a must-have for successful and reliable operations. This includes which part of the system needs to be updated, what maintenance activities need to be conducted and how often, prioritizing the maintenance tasks, and determining the resources required.

2. Make sure you have a regular system backup plan in place and have an updated backup before any maintenance work. This is important to bring back the system to its original state in case of any work mishaps.

3. Test any planned work before actually applying the update or the change. Test the change thoroughly in a lab or staging environment to ensure it is functioning as expected.

4. Monitor the system after the updates have been deployed. This includes monitoring performance metrics, running automated checks, checking users' feedback, and checking system logs.

5. Update your documentation with the changes, and document the maintenance and update outcome. The documentation will help ensure that the maintenance and update process is repeatable for future reference, and help troubleshoot in case of any issues that may happen in the future.

In the beginning, the maintenance plan may not be as perfect as you may like it to be, but as the process is repeated during the lifetime of the system, the process will eventually get refined to a very accurate level.

User satisfaction

Monitoring and improving user satisfaction are sometimes underestimated in the success of deploying any new IT system. By gathering feedback from the system's internal and external customers, analyzing that feedback, prioritizing changes, implementing changes, monitoring progress, and continuously improving, we can ensure that the system meets customer requirements.

The following is a four-step cycle for ensuring high customer satisfaction post-deployment:

Figure 12.5: The four-step customer satisfaction cycle

1. The first step in monitoring our customer satisfaction is to gather feedback from the system users. The feedback can be collected through surveys, direct customer interaction and visits, phone conversations, and so on.

2. Analyze the gathered feedback to identify common complaints, common patterns, and specific use cases that may have not been covered during the system testing phase. This will help us understand the strengths of the system and the areas where we need to improve.

3. Prioritize whatever system changes are decided as an outcome of the gathered feedback. This part should have the biggest impact on customer satisfaction. It will show customers that you are addressing their concerns, reacting to their requests, and, sometimes, even being proactive to customer needs.

4. Implement the changes as prioritized. Start with the changes with the highest impact and lowest effort similar to what we discussed in *Chapter 11* under the *Risk mitigation* section and in *Figure 11.2*.

We need to continuously regather customer feedback to regularly monitor customer satisfaction progress. This will ensure that the changes being carried out are having the desired customer satisfaction effect.

The four-step process will help continuously meet customer needs and improve customer satisfaction accordingly. The process helps also constantly improve the system features, supportability, reliability, and stability to enhance the overall user experience.

In this section, we covered the post-deployment review process, the different aspects, and activities that need to be considered when conducting the review, and how that is essential in the overall success of the ABC-Intelligent-MSA system operations.

Summary

In this chapter, we discussed, the various steps involved in the successful deployment of the new system. We talked about the importance of overcoming the system deployment dependencies, the importance of building structured test cases, and the steps involved in testing and tuning the system.

By following the steps outlined in this chapter, organizations can ensure the successful deployment of their MSA system, including overcoming dependencies, integrating with the monolith during the transition phase, testing and tuning the system, and conducting a post-deployment review. The chapter concluded by emphasizing the significance of following a customer satisfaction cycle and having customers engaged in the process of adapting the new system's operations.

Index

`Packtpub.com`

Subscribe to our online digital library for full access to over 7,000 books and videos, as well as industry leading tools to help you plan your personal development and advance your career. For more information, please visit our website.

Why subscribe?

- Spend less time learning and more time coding with practical eBooks and Videos from over 4,000 industry professionals

- Improve your learning with Skill Plans built especially for you

- Get a free eBook or video every month

- Fully searchable for easy access to vital information

- Copy and paste, print, and bookmark content

Did you know that Packt offers eBook versions of every book published, with PDF and ePub files available? You can upgrade to the eBook version at `packtpub.com` and as a print book customer, you are entitled to a discount on the eBook copy. Get in touch with us at `customercare@packtpub.com` for more details.

At `www.packtpub.com`, you can also read a collection of free technical articles, sign up for a range of free newsletters, and receive exclusive discounts and offers on Packt books and eBooks.

Other Books You May Enjoy

If you enjoyed this book, you may be interested in these other books by Packt:

Applying Math with Python - Second Edition

Sam Morley

ISBN: 9781804618370

- Become familiar with basic Python packages, tools, and libraries for solving mathematical problems

- Explore real-world applications of mathematics to reduce a problem in optimization

- Understand the core concepts of applied mathematics and their application in computer science

- Find out how to choose the most suitable package, tool, or technique to solve a problem

- Implement basic mathematical plotting, change plot styles, and add labels to plots using Matplotlib

- Get to grips with probability theory with the Bayesian inference and Markov Chain Monte Carlo (MCMC) methods

Applied Machine Learning Explainability Techniques

Aditya Bhattacharya

ISBN: 9781803246154

- Explore various explanation methods and their evaluation criteria
- Learn model explanation methods for structured and unstructured data
- Apply data-centric XAI for practical problem-solving
- Hands-on exposure to LIME, SHAP, TCAV, DALEX, ALIBI, DiCE, and others
- Discover industrial best practices for explainable ML systems
- Use user-centric XAI to bring AI closer to non-technical end users
- Address open challenges in XAI using the recommended guidelines

Packt is searching for authors like you

If you're interested in becoming an author for Packt, please visit `authors.packtpub.com` and apply today. We have worked with thousands of developers and tech professionals, just like you, to help them share their insight with the global tech community. You can make a general application, apply for a specific hot topic that we are recruiting an author for, or submit your own idea.

Share Your Thoughts

Now you've finished *Machine Learning in Microservices*, we'd love to hear your thoughts! Scan the QR code below to go straight to the Amazon review page for this book and share your feedback or leave a review on the site that you purchased it from.

`https://packt.link/r/1-804-61774-1`

Your review is important to us and the tech community and will help us make sure we're delivering excellent quality content.

Download a free PDF copy of this book

Thanks for purchasing this book!

Do you like to read on the go but are unable to carry your print books everywhere? Is your eBook purchase not compatible with the device of your choice?

Don't worry, now with every Packt book you get a DRM-free PDF version of that book at no cost.

Read anywhere, any place, on any device. Search, copy, and paste code from your favorite technical books directly into your application.

The perks don't stop there, you can get exclusive access to discounts, newsletters, and great free content in your inbox daily

Follow these simple steps to get the benefits:

1. Scan the QR code or visit the link below

https://packt.link/free-ebook/9781804617748

2. Submit your proof of purchase
3. That's it! We'll send your free PDF and other benefits to your email directly

Download a free PDF copy of this book

www.ingramcontent.com/pod-product-compliance
Lightning Source LLC
Chambersburg PA
CBHW060532060326
40690CB00017B/3462

* 9 7 8 1 8 0 4 6 1 7 7 4 8 *